T0276059

SpringerBriefs in Mathematics

Series Editors

Nicola Bellomo
Michele Benzi
Palle E. T. Jorgensen
Tatsien Li
Roderick Melnik
Otmar Scherzer
Benjamin Steinberg
Lothar Reichel
Yuri Tschinkel
G. George Yin
Ping Zhang

SpringerBriefs in Mathematics showcases expositions in all areas of mathematics and applied mathematics. Manuscripts presenting new results or a single new result in a classical field, new field, or an emerging topic, applications, or bridges between new results and already published works, are encouraged. The series is intended for mathematicians and applied mathematicians.

More information about this series at http://www.springer.com/series/10030

Shravan Hanasoge

Imaging Convection
and Magnetism in the Sun

 Springer

Shravan Hanasoge
Department of Astronomy and Astrophysics
Tata Institute of Fundamental Research
Mumbai, India

ISSN 2191-8198   ISSN 2191-8201   (electronic)
SpringerBriefs in Mathematics
ISBN 978-3-319-27328-0   ISBN 978-3-319-27330-3   (eBook)
DOI 10.1007/978-3-319-27330-3

Library of Congress Control Number: 2015957966

Mathematics Subject Classification (2010): 35B27, 47A05, 47B25, 15A29, 15A09, 65N12, 65N06, 65N21, 35R30, 65N22, 65N80, 35L05, 35L10, 35L15, 35L20, 35Q93, 85A30, 76W05, 76R10, 78M20, 78M22, 86A15, 86A22, 35Q31, 35Q35

Springer Cham Heidelberg New York Dordrecht London

Printed on acid-free paper

Springer International Publishing AG Switzerland is part of Springer Science+Business Media (www.springer.com)

For my parents.

Foreword

Helioseismology revolutionized the study of the Sun. As recently as the early 1970s, it seemed inconceivable that scientists could study the solar interior observationally. All understanding of the interior structure and dynamics of the Sun was based on observations of the surface properties of the Sun and other stars and theoretical modelling of stellar structure and evolution.

Much of the revolution that has occurred in the intervening time has been based on global helioseismology: the study of the solar interior using the observed properties of global resonant oscillations of the Sun. The observational data are Doppler velocity measurements and brightness variations at or just above the Sun's photosphere (its visible surface). The subject leapt forward with new, highly spatially resolved data from ground-based networks and from space. These highly resolved measurements in turn opened up new areas of helioseismology, collectively known as local helioseismology, that are based on the analysis of properties of waves observed on the solar surface. Dr. Hanasoge has been a pioneer in the modelling and analysis of such waves and has done much to bridge the gap between two disciplines: the seismology of the Earth and the seismology of the Sun.

In this book, Dr. Hanasoge presents a comprehensive treatment of the forward and inverse modelling of helioseismology from a fundamental wave theoretic perspective. His treatment complements but goes far beyond the other textbooks currently available that treat helioseismology. Drawing on an extensive literature and experience in geo-seismology, Dr. Hanasoge demonstrates how these approaches can be applied to the seismic study of the Sun. This book is a rich resource for any student or researcher wishing to understand or develop local helioseismic techniques for measuring the internal structure, dynamics or magnetic fields in the Sun. As such, I expect it to become widely used in the helioseismic research community. Some of the material is hard, to be sure. But Dr. Hanasoge's clear style and expertise will open up the field to the diligent student.

The field of helioseismology has much still to offer for our understanding of the physics of the solar interior. Researchers who wish to engage with this challenging but intriguing field will find much to value in this book.

Boulder, CO, USA   Michael Thompson
March 2015

Preface

Our pedestrian Sun is an extraordinary object, exhibiting complex dynamics; we forget at times that we live with a star. The first historical record of the inconstancy of the Sun is attributed to the Chinese, who, more than two millennia ago, made sketches of a spotted Sun. Fast forward to the early sixteen hundreds Galileo, among other contemporaries, began to observe and describe these spots with his then-newly invented telescope. He was the first to demonstrate they were on the surface of the Sun, implying that they were indeed solar in origin. Four hundred years on, contemporary solar physics is driven by expensive international space missions and billion dollar space telescopes observe the Sun at high resolution, beaming down a terabyte of data *every day*. It is striking to view the progression of our understanding of the Sun, from ancient sketches of sunspots to detailed models of structure and internal rotation and elaborate theories of global magnetic field generation. The refinement in our appreciation of the structure and dynamics of the Sun from what we knew even in the late nineteen sixties is remarkable. Not only has the Sun shed insight on stellar physics, i.e. the study of the structure, evolution and dynamics of stars, it continues to be a relevant object of study today.

The Sun's magnetic field directly affects Earth climate and space weather. The internal mechanism that governs the Sun's large-scale field reversals is a subject of great interest to dynamo theorists and astrophysicists. The properties of convection in the highly stratified environment of the solar convection zone provide insights in a parameter regime inaccessible to computation and laboratory experiment and are therefore of relevance to fluid mechanicians. Solar and stellar physics are very important branches of astronomy and astrophysics, whose study is enabled by substantial financial support from international space agencies. There have been some ten satellite missions in the fields of solar and stellar physics and the installation of a number of ground-based instruments over the past two decades, representing a substantial investment by the scientific community. As a consequence, high-quality observations of the Sun and stars are now abundantly available and so the burning questions now almost entirely concern accurate interpretation. It is in this backdrop that this monograph finds its relevance.

The interior of the Sun is opaque and therefore not amenable to optical imaging. The discovery that acoustic waves propagate within the solar interior and appear at the surface, where they can be directly observed (optically), opened up an exciting field of study in the nineteen sixties: helioseismology, the study of the oscillations of the Sun as a means of inferring its internal structure and dynamics. From the incipient struggle to take accurate observations of the oscillations of the Sun to interpreting the fine structure of its resonant modes and eventually inferring the internal structure and dynamics, helioseismology has come a long way. The contemporary focus is on the development and implementation of techniques to create 3-D images of convection and magnetism in the solar interior. This monograph attempts to introduce the latest computational and theoretical methods to the interested reader. Some proficiency in basic numerical methods, differential equations and linear algebra is a requisite to appreciate the material presented here.

Mumbai, India   Shravan Hanasoge
September 2015

Acknowledgements

I would like to thank Katepalli Sreenivasan for his encouragement and help towards completing this monograph. The idea for this endeavour was seeded over a discussion with Andreas Fichtner. Tom Duvall kindly provided me an HMI power spectrum and time-distance diagram. I thank H S Mukunda, H M Antia, Kuldeep Verma and Jishnu Bhattacharya for their comments and feedback.

Contents

Contents

xv

Acronyms

FWI   Full waveform inversion, a seismic technique that attempts to match the full observed waveform

HMI   Helioseismic and magnetic imager, an instrument onboard SDO that takes high-resolution Doppler images of the photosphere

MDI   Michelson Doppler imager, an instrument onboard SOHO that took measurements of the Doppler velocity of the photosphere of the Sun

MHD   Magnetohydrodynamics, which deals with the behaviour of magnetic fluids

SDO   Solar dynamics observatory, a satellite launched by NASA in February 2010 to observe the Sun

SOHO   Solar and heliospheric observatory, a satellite, built by NASA and ESA, that observed the Sun between 1996 and April 2011

Chapter 1
Introduction

The Sun was formed roughly 4.5 billion years ago from the gravitational collapse of a gaseous cloud. The core of this cloud settled in the center, gathering an overwhelming fraction of the matter of the cloud ($\sim$ 99.85%) while the rest formed a disk that would eventually turn into the solar system. Owing to gravitational forcing, the central mass gained temperature and the core, reaching a critical temperature of about 15 million K, began to undergo nuclear fusion. The present Sun is primarily composed of Hydrogen ($\sim$ 71%) and Helium ($\sim$ 27%), the rest being metals (in astrophysical jargon, elements other than Hydrogen and Helium are termed 'metals').

Our Sun is a main-sequence star, classified as a spectral type G2V based on the wavelength at which its radiation peaks (yellow as seen from Earth, implying a surface temperature of about 5770 K). Through observations of planetary motion, the solar mass can be measured to a high degree of accuracy (limited only by errors in the knowledge of the gravitational constant G). The Sun's luminosity, estimated to be 3.846×10^{26} W, is powered by nuclear fusion in the core. Roughly 600 million metric tonnes of Hydrogen are transformed every second to Helium in the hot, dense solar core to sustain one solar luminosity. A schematic of the Sun's inner layers and atmosphere is shown in Figure 1.1.

The Sun, like many of its main-sequence cousins, exhibits large-scale cyclic magnetic activity, with its global magnetic field reversing polarity every eleven years. A prominent manifestation of solar magnetism is the regular appearance of sunspots (see Figure 1.2), disc-like structures as seen at the surface of the Sun. Sunspots emerge at different latitudes at different phases of the solar cycle, shown in the 'butterfly' diagram in Figure 1.3. Early on in the cycle, sunspots emerge at high latitudes and as the cycle progresses, they appear closer to the equator. Eventually the cycle ends (solar minimum) and another begins.

Life on Earth is directly affected by the magnetic variability of the Sun. During each magnetic cycle, the rising and waning phases are causally linked to irradiance variations in ultraviolet emission, which in turn forces Earth climate. Solar magnetic phenomena can result in the ejection of high-energy particles, which in turn enter

© The Author 2015
S. Hanasoge, *Imaging Convection and Magnetism in the Sun*, SpringerBriefs in Mathematics, DOI 10.1007/978-3-319-27330-3_1

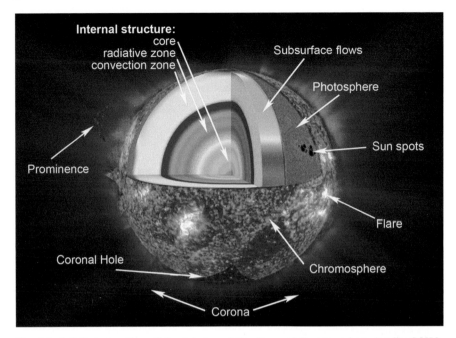

Fig. 1.1 Artist's conception of the interior of the Sun and its atmosphere (credit: SOHO, ESA/NASA). The Sun can be optically imaged only from the photosphere and outwards. Nuclear fusion transforms Hydrogen in the core to Helium, releasing a solar luminosity's worth of heat flux (3.846×10^{26} W). This heat is transported by radiation to around 0.7 $R_{\odot}$, beyond which convection takes over as the dominant heat-transport mechanism. Finally thermal energy is released into space at the photosphere by free-streaming radiation. The other layers of the Sun exhibit violent magnetically driven eruptions, sunspots, shocks, and vigorous convection. Helioseismology is the primary means by which we can image layers deeper than the photosphere.

and alter Earth's atmosphere. The climate reacts to these variations in complex, non-linear ways (Haigh 2007). Further, space instrumentation and telecommunication are susceptible to solar high-energy eruptive events (Schrijver and Zwaan 2000; Pulkkinen 2007).

1.1 Thermal transport

The Sun serves as an astrophysical benchmark, contributing to the development of our understanding of stellar evolution, stellar interiors, coronae. The transport of heat from fusion in the cores of stars to their exterior layers is primarily accomplished by a combination of radiation and convection while conduction is ineffective. Convection is a macroscopic phenomenon, where the transport of heat is accomplished by the direct movement of the fluid. For instance, in a pan of water heated from below, hot fluid at the bottom becomes buoyant and rises, transferring

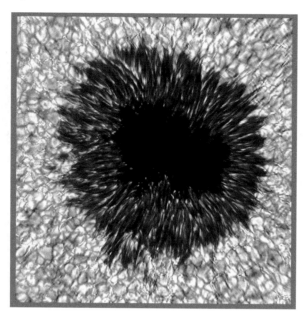

Fig. 1.2 A sunspot sitting amid a field of granules (convective structures in the near-surface layers) observed in intensity by the Sacramento Peak Observatory at the National Solar Observatory (Credit: National Solar Observatory). Sunspots are non-axisymmetric and may exhibit rotation, magnetic twist, etc. The typical sunspot is characterized by a central cool (as much as 1000 K lower than the ambient temperature) and magnetically strong umbra ($\sim$ 3000G) surrounded by a complex fibrous magnetic penumbra. The sunspot shown here has a width of roughly 50 Mm. The surrounding granules are in a state of vigorous convection, comprising upwelling hot fluid in the centers around which lanes of cool fluid are falling back into the interior. Helioseismology provides a powerful means of examining the interior structure and dynamics of sunspots.

heat to its locally surrounding cooler (upper) layers. In contrast, conduction is a microscopic phenomenon, occurring due to the repeated scattering of particles. Higher temperatures imply larger particle kinetic energies, and random collisions result in the transfer of particle kinetic energy; thus the transport is diffusive. Similarly, radiation is also a diffusive phenomenon: energetic photons transport the heat, undergoing repeated absorption and re-emission by particles in the stellar interior. The degree to which radiation or conduction is successful in removing heat is determined by the *mean free path $\bar{l}$* of photons or particles, i.e. the typical distance between consecutive interactions or collisions. Generally, in stellar interiors, particles possess much shorter mean free paths than photons, and conduction plays an insignificant role. The balance between radiation and convection is set by the local temperature gradient; if it exceeds some critical threshold (and the photon mean free path becomes sufficiently short), convective instabilities set in (Böhm-Vitense 1958). The transport of heat from the core to about 70% of the radius of the Sun ($R_\odot = 696,000$ km) is accomplished by radiation (see Figure 1.1). Photons diffuse outwards, undergoing repeated free-free scattering, which occurs when

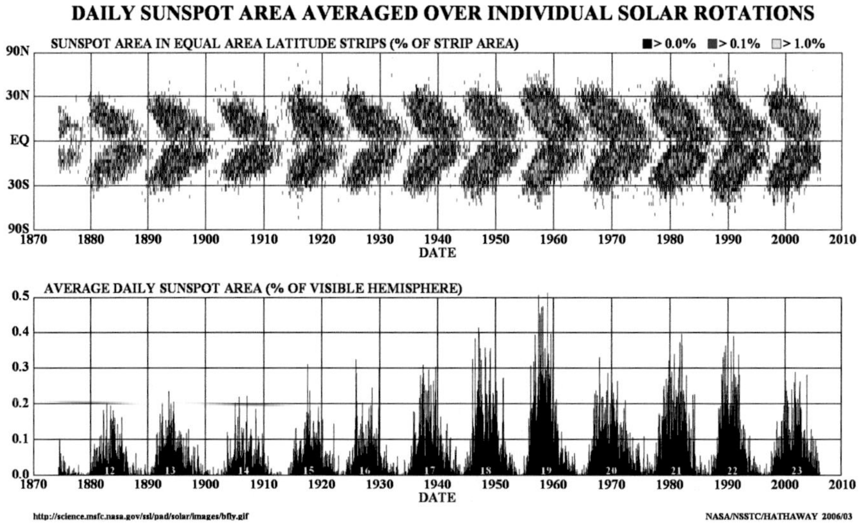

DAILY SUNSPOT AREA AVERAGED OVER INDIVIDUAL SOLAR ROTATIONS

Fig. 1.3 Evolution of the Sun's magnetism over the past century (Credit: NASA). The butterfly diagram, as it is termed, involves placing a dot on the corresponding latitude whose color indicates the area occupied by sunspots and plotting this as a function of time. The butterfly diagram indicates cyclic magnetic activity on a timescale of around 11 years, growing from a state of low solar activity (solar minimum) to reaching a peak in activity (solar maximum). At the start of the cycle, sunspots emerge preferentially at high latitudes (30 degrees) with the emergence locations shifting closer to the equator (in a statistical sense) as the cycle progresses. The plot below tracks the sunspot area as a function of time, showing a secular change in the amplitude of the cycle.

photons are absorbed and emitted by electrons in the highly ionized solar plasma (Stix 2004). The frequency-dependent Rosseland-mean *opacity* $\kappa_\nu = 1/(\rho \bar{l})_\nu$, where ρ is the density of the medium, ν the frequency of the photons, and $\bar{l}_\nu$ the frequency-dependent mean free path, is a measure of the freedom with which photons can propagate.

The density of the Sun decreases with increasing radius, as does the temperature, resulting in a direct increase in the opacity of solar plasma. At the outer edge of the radiative zone, heat transport can no longer be sustained along the local (radially directed) temperature gradient by radiation alone. Convective instabilities form, resulting in the onset of convection. The outer 30% of the Sun is termed the convection zone, for the reason that heat transport is accomplished by motions of plasma. In the near-surface layers of the Sun, the density falls very rapidly with increasing radius: the density scale height, a measure of how rapidly density is varying, $H_\rho = -(d\ln\rho/dr)^{-1}$, where r is the radial distance from the center of the Sun, is on the order of a few hundred kms. Opacity in the convection zone is high, so the solar interior is optically inaccessible. Close to the surface, the density falls so rapidly and the plasma becomes sufficiently rarefied that the photon mean free path becomes very large (on the order of 100 km; Judge et al 2014). At this surface layer, termed the photosphere, heat ceases to be transported by convection and

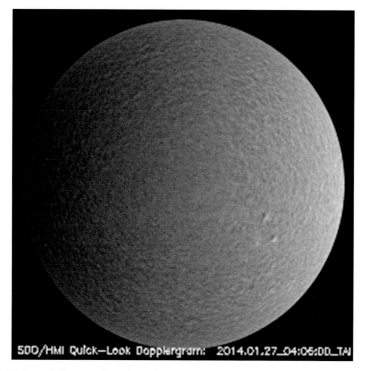

Fig. 1.4 High-resolution Doppler velocity image of the photosphere of the Sun taken by the Helioseismic and Magnetic Imager onboard the Solar Dynamics Observatory satellite. A charge-couple device takes filtered snapshots of the photospheric velocity field at a resolution of 4096 × 4096 pixels every 45 seconds. Solar rotation has not been subtracted, creating an asymmetry between the east and west limbs.

free-streaming radiation takes over. Direct observations of the Sun begin at the photosphere, as depicted in Figure 1.4. The photosphere shows a near-black-body spectrum of radiation and a range of absorption lines are apparent. Instruments can be designed to observe absorption lines formed at the photosphere and to measure intensity variations. Motions along the line of sight (in relation to the observer) that buffet layers of the photosphere cause the frequency at which these absorption lines are formed to be Doppler shifted with respect to the observer. Accurately measuring these Doppler shifts therefore allows for inferring the velocity field at the surface. Thus far, satellites from a single point of view have been imaging the photosphere in this manner, and thus the velocity can only be observed from the line of sight of that instrument.

Granules, Rayleigh-Bénard-type convective structures (see Figure 1.2), transport heat in the immediate sub-surface to the photosphere of the Sun. Granules are on the order of 1000–2000 km in size, exhibiting supersonic downflows, i.e., overturning convective flows, which in turn excite acoustic waves (see the review by, e.g., Nordlund et al 2009). The efficiency with which high-speed flows channel energy into acoustic waves is known to vary as M^8, where M is the Mach number (Lighthill 1952). Thus even mildly supersonic flows can very effectively force acoustic waves. It is widely thought that this mechanism is the source of solar oscillations. Numerous studies (e.g. Kumar and Basu 2000) have shown that various properties of solar oscillations are well reproduced by sources placed very close to the photosphere.

Problems in solar physics do not easily give in to resolution, particularly those pertaining to the atmospheric layers of the Sun. For instance, in the solar chromosphere and corona, densities are so low that local thermodynamic equilibrium (whence the notion of a temperature emerges) is not valid. Complex non-linearities arising from magnetic reconnection and supersonic turbulence amid highly locally stratified plasma are pervasive. Among the problems in solar physics, helioseismology is the simplest, best defined, and the cleanest. Helioseismology (see Christensen-Dalsgaard 2002; Gizon and Birch 2005; Gizon et al 2010; for reviews) is a collection of methods applied to infer the interior structure and dynamics of the Sun through the interpretation of surface observations of its oscillations.

1.2 Seismology

The definitive discovery of oscillations of the Sun by Leighton et al (1962) opened up the possibility of seismically imaging the optically inaccessible interior of the Sun. The authors recognized, strikingly, the potential for probing the surface layers of the Sun but they could little have imagined the myriad ways in which helioseismology has evolved. Observations by Deubner (1975) and Rhodes et al (1977) confirmed the global nature of solar oscillations. Simultaneously, theoretical progress led to the interpretation of the structure of the normal modes of the Sun, identifying resonance criteria that would result in sets of acoustic modes (Leibacher and Stein 1971). A thorough exposition on the properties of solar and stellar oscillations is given in Christensen–Dalsgaard (2003). Figure 1.5 shows the power spectrum of the photospheric line-of-sight velocity field of the Sun recorded by the Michelson Doppler Imager (MDI; Scherrer et al 1995) onboard the Solar and Heliospheric Observatory launched by the European Space Agency in 1995. MDI took snapshots of the velocity field as a function of latitude and longitude (measured using Doppler imaging) once a minute. This spatio-temporal sequence is projected onto spherical harmonics (note that the far side of the Sun is not observed so the projection on to spherical-harmonic space is imperfect) and subsequently Fourier transformed in time. The squared absolute value of this transformed data cube is summed over azimuthal order m, to compute the power spectrum. It is seen in Figure 1.5 that power is concentrated along a series of ridges, each of which, upon closer examination, is seen to comprise a large collection of resonant modes of the Sun. The so-called f 'fundamental' surface-gravity mode is the farthest right of ridges, while the rest are acoustic p 'pressure' modes. Much as an oscillating string tied at both ends supports a discrete set of modes, termed normal modes, the Sun pulsates in specific ways. This particular manner of oscillation is dictated by the structure and dynamics of the interior and thus the frequencies of oscillation of the Sun provide insight into its structure. Resonant modes are by definition 'trapped' in that they form because of boundary conditions in finite spatial region, termed a 'cavity' (as opposed to a traveling wave which propagates in an unrestricted fashion). Thus the way in which modal power concentrates serves as a proxy for probing the structure of the resonance cavity. Properties of resonance cavities directly relate to structure, composition, and dynamics of the interior of the Sun.

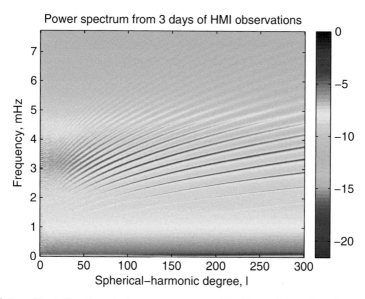

Fig. 1.5 Logarithmic Doppler velocity power spectrum of the Sun as observed by the HMI instrument. The horizontal axis is the spherical-harmonic degree, ℓ. This is related to the wavelength by the relation $\lambda = 2\pi R_{\odot}/\sqrt{\ell(\ell+1)}$. Note that HMI observes the Sun at high resolution so the horizontal axis can be extended for ℓ up to 3000. The vertical axis is the frequency expressed in milli Hertz. Red indicates high power. At frequencies below 1 mHz, the power is due to convection. Above 1 mHz, power is concentrated along a series of ridges which comprise a discrete (closely spaced) set of modes (Courtesy: Tom Duvall).

Seismology is a widely used means to image the hidden interiors of opaque bodies. Seismology of Earth, for instance, involves measuring ground displacements (at Earth's surface) due to waves generated by earthquakes or by other means (e.g., ocean waves or by manmade explosions). In the Sun and other Sun-like stars, vigorous surface convection excites waves. Seismology occasionally provides dramatically new information about the interior; for the most part however, it is a powerful technique for providing fresh constraints on refining the physics of pre-existing models and reducing uncertainties relating to models of stellar structure. The first step towards posing a seismology problem is to characterize the physics of wave propagation in that system. In the case of the Sun, the model of small-amplitude (linear) waves propagating in a stratified environment is very successful in capturing mode physics (e.g., Christensen-Dalsgaard 2002). This is known as a *forward* model, in that given the structure of the Sun, it outputs the predicted wavefield. These predictions are then compared with observations and a misfit function that captures the difference between the two is constructed. The *inverse* problem then asks the question: what are the best-fit structural properties of the Sun that minimize the misfit subject to the constraint that the forward model be satisfied. The operator that governs linear wave propagation possesses eigenvalues and eigenfunctions. Since the medium is being sensed by waves, the properties must necessarily be expressible in the basis comprising eigenfunctions of the operator. Thus the mathematical problem of seismology lies in determining that projection. Whether there exists such a projection for the given (limited) set of seismic measurements, that will also likely be corrupted by noise, is an important issue. A limited set of measurements suggests that the basis might not be complete and it may therefore not be possible to find either a unique or accurate solution to the seismic inverse problem. Noise can have a similar effect.

Wave propagation in the Sun is extremely well described as linear, i.e., small-amplitude disturbances in comparison to the local sound speed. Further, the normal modes of the Sun are trapped resonances in cavities that lie within the interior, making this a classic seismology problem. Thus one can construct a linear Green's function approach to relate surface wavefield measurements to interior properties of the Sun. For these reasons, successes in helioseismology have been numerous, contributing substantially to the greater body of knowledge in stellar physics. Subsequent to the identification and classification of solar oscillation modes (Leibacher and Stein 1971; Rhodes et al 1977; Duvall 1982), early efforts in helioseismology centered around accurately constraining the internal structure and rotation of the Sun. We describe two related highly influential results here.

1.2.1 Structure Inversions and the Solar Neutrino Problem

A seminal result to emerge from helioseismology was the precise recovery of the solar interior sound-speed profile (Christensen-Dalsgaard et al 1996), shown in Figure 1.6. Models of solar structure obtained from stellar evolution codes predicted oscillation frequencies that differed from observations. This led to improvements in stellar modeling such as the inclusion of gravitational settling (Proffitt and Michaud 1990; Thoul et al 1994), which in turn resulted in better agreement between observed and predicted oscillation frequencies (see also Figure 1.7). The core temperature of the Sun as predicted by standard solar models suggested that nuclear fusion via the p-p chain was the dominant mechanism of heat generation. Neutrino-detection experiments that had been launched ever since 1968 saw around a third of the predicted flux, and doubt had been directed towards whether models of solar structure were accurate. However, with the advent of sophisticated experiments such as Super Kamiokande and the Sudbury Neutrino Observatory (SNO), the theoretically anticipated flavor oscillations were finally detected. Neutrinos exist in three flavors, alternating between them and early experiments were capable of detecting only one flavor. Super Kamiokande and SNO designed specifically to test the flavor-oscillation hypothesis, successfully discovered the existence of Neutrinos in alternate flavors. For their work on neutrino detection, Ray Davis and Masatoshi Koshiba were awarded the Nobel prize in 2003.

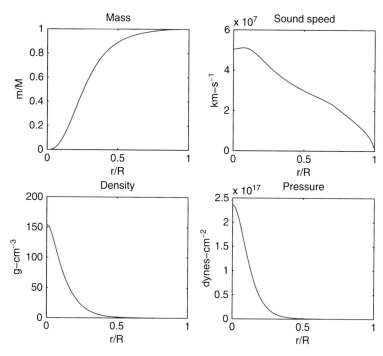

Fig. 1.6 Structure model of the Sun (model S Christensen-Dalsgaard et al 1996). The radius of the Sun is $R_\odot = 696 \times 10^6$ m and mass $M_\odot = 1.98 \times 10^{33}$ g. Density and pressure drop steeply in the near-surface layers.

1.2.2 Internal Rotation

So-called standard models of the Sun are one dimensional, representing spherically symmetric properties (such as density, sound speed, gravity, pressure, etc.) as a function radius. Helioseismology has the capability to go beyond these standard models, and the inference of internal rotation as a function of latitude and radius is a high-fidelity result that has withstood repeated testing. Typically structural properties of stars are derived using stellar evolution codes, which involve tracking the evolution of a star from its origin as a molecular cloud to the start of fusion in the core, and its eventual development. Because of the vast timescales (billions of years) that evolution takes, numerical models of stellar development attempt the solution

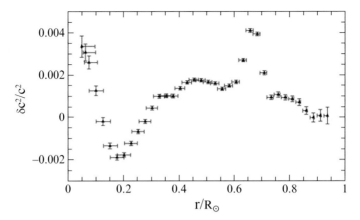

Fig. 1.7 Differences between models of sound-speed variation in the Sun obtained from seismic inference and from stellar evolution calculations (δc^2). The models match very closely (e.g., Christensen-Dalsgaard 2002).

of a reduced system of equations that generally do not include rotation (e.g., see, Christensen–Dalsgaard 2008). Thus rotation must be measured directly from oscillation frequencies. Oscillation modes are typically characterized in terms of spherical harmonics and radial orders. Each mode is uniquely identified by a harmonic degree ℓ, azimuthal order m, and radial order n, where $|m| \leq \ell$. For a quiescent non-rotating sphere, modal symmetry ensures that $\omega_{\ell mn} = \omega_{\ell 0n}$ where $\omega_{\ell mn}$ is the oscillation frequency of mode (ℓ, m, n). Rotation lifts this degeneracy, inducing the shifts, $\omega_{\ell mn} = \omega_{\ell 0n} + m\Omega$ and $\omega_{\ell,-m,n} = \omega_{\ell 0n} - m\Omega$, where Ω is a solid-body rotation rate (for weak rotation $\Omega \ll \omega_{\ell 0}$). Frequency differences between the $\pm m$ modes can thus be directly related to rotation. For differential rotation, i.e. $\Omega = \Omega(\theta, r)$, where θ is latitude and r is radial distance from the center, shifts in frequency are related to mode-energy-weighted integrals of the rotation rate (in some rough sense). Thus the analysis is more involved and an inverse problem needs be solved but the principle is the same. In this manner, the full set of measured frequencies can be leveraged to estimate the internal rotation rate of the Sun. The important results of, e.g., Schou et al (1998) showed that the outer convective envelope rotates differentially, equatorial layers exhibiting a faster rotation rate than the poles. Contours of constant rotation as inferred from observations taken by MDI are shown in Figure 1.8. The Taylor-Proudman theorem, that applies to rotating spheres of fluid, suggests that rotation contours should be parallel to the axis and yet, solar rotation is nearly solid body with contours that are radially directed. How such rotation contours are formed and maintained continues to be a matter of some debate and is a counter-intuitive result (Kitchatinov and Rüdiger 2005; Rempel 2005; Balbus 2009; Miesch and Hindman 2011; Balbus and Schaan 2012; Hanasoge et al 2012b; Miesch et al 2012).

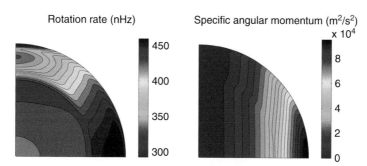

Fig. 1.8 Helioseismically inferred internal rotation of the Sun (Ω, left panel) and specific angular momentum ($= \Omega^2 r^2 \sin^2 \theta$, where r is radius and θ is co-latitude, right panel). Resonant-mode frequencies, measured from MDI observations, were used to make these inferences. The iso-rotation contours are not parallel to the axis, as Taylor-Proudman balance might suggest. Courtesy: H. M. Antia.

1.2.3 Non-axisymmetry

To leading order, the Sun is an axisymmetric body, where structure and rotation are only functions of radius and latitude. However images such as Figure 1.2 present strong evidence that critical to gaining an appreciation of solar dynamics rests on exploring phenomena that are three-dimensional, non-axisymmetric, and time-varying. Sunspots (Figure 1.2), localized thermal hotspots, convective systems, magnetic fields in the interior are topics of great interest in helioseismology. Localized hotspots in the interior (termed thermal asphericities) have long been conjectured and their faint signatures have been observed in seismic measurements (e.g., Swisdak and Zweibel 1999).

Contemporary helioseismology has turned to addressing a grand challenge in solar physics: how are global magnetic fields generated and sustained? The solar dynamo likely relies on the differential rotation of the convective envelope, meridional circulation, the radiative stability of the interior, and the transition layer between radiative and convective zones, all of which are active research areas in and of themselves. It is increasingly clear that these phenomena are likely acting together (in some sense) to support the dynamo (a fluid that generates and sustains its own magnetic field), a process we now know to be prevalent in a broad swath of main-sequence Sun-like stars (Noyes et al 1984; Brandenburg 2005). A number of unknowns have made the identification of the underlying mechanisms harder, such as imaging the equator-to-pole (large-scale) meridional circulation. Where are sunspots rooted? What is the nature of interior convection? The resolution of what drives the dynamo will thus require a greater understanding of large-scale fluid phenomena in the Sun.

1.3 Seismic Observations of the Sun

In the photospheric layers, density decreases rapidly and the photon mean free path becomes very large. Thus direct optical imaging of the photosphere is feasible. For all practical purposes, the photosphere can be treated as a black-body emitter, and the dominant wavelengths of photons may be easily estimated based on the temperature, assuming local thermodynamic equilibrium (despite the low densities prevalent in the photosphere). Absorption lines form at the photosphere which is the deepest visible layer of the Sun. Observing these absorption lines allows for characterizing properties of the surface layers. For instance, lines are Doppler shifted with respect to the observer due to motions of the plasma, such as buffeting by waves, rotation, and convection. Measurements of Doppler shifts of absorption lines can be used to recover the surface velocity field (not the vector velocity, just the component projected along the line-of-sight direction). Further, magnetic field causes absorption lines to undergo a Zeeman broadening, which in turn can be used to infer the field. Unlike Doppler velocity imaging where only the line-of-sight projection is measured, the full vector magnetic field can be obtained using a combination of Zeeman broadening and Stokes-parameter measurements (Borrero et al 2007).

Prior to the modern era, there were several campaigns that served critical roles in progress on helioseismology (e.g., Ulrich 1970; Leibacher and Stein 1971; Rhodes et al 1977). Some of the most important observational sequences, in terms of the impact their analyses led to, were taken during the Antarctic austral summer. Continuous solar visibility in conjunction with excellent seeing conditions due to pristine atmospheric conditions allowed for taking high-quality observations. These measurements led to advances in methodology (Duvall et al 1993), seismic inferences (Duvall et al 1986, 1996), and in the characterization of photospheric convection (Schrijver et al 1997).

A number of observational efforts to image the solar photosphere have been made but the modern era of continuous seismic surveys of the Sun began with the Global Oscillation Network Group (GONG), which was a group of 6 telescopes stationed in different time zones to achieve 24-hour coverage (Harvey et al 1988; Hill et al 1994).

The age of space-based observation arrived with the launch of Solar and Heliospheric Observatory (SOHO) in 1996. SOHO was placed on the Lagrange (L1) orbit, where the orbital revolution period around the Sun is identical to that of Earth's. Three instruments were placed onboard, one of which was the Michelson Doppler Imager (MDI; Scherrer et al 1995). MDI had a charge-coupled device camera (CCD) that took images of the Sun every 60 seconds, at a resolution of 1024×1024 pixels and in specific instances, high-resolution images of parts of the visible disk. Radiation from the Sun was filtered by MDI sequentially, eventually passing through a tunable pair of Michelson interferometers, allowing for the measurement and recording of filtergrams around the Nickel (Ni I, 6768Å) line. SOHO as a mission achieved great success, whose observations led to a significant improvement in our understanding of solar irradiance, the seismic wavefield, magnetic field measurements of the photosphere, etc. MDI was finally shut off in April 2011, significantly outliving its planned lifetime and providing a nearly unbroken stretch of 16 years of observations. In 2010, the successor to SOHO, the Solar Dynamics Observatory (SDO) was launched. SDO is in geosynchronous orbit around Earth, and contains three instruments that are in excellent health at the time of this writing. The Helioseismic and Magnetic Imager (HMI; Schou et al 2012) takes 4096×4096-pixel snapshots of the photosphere every 45 seconds. HMI observes the Sun using the Iron Fe I 6173Å absorption line. Among the data products are the line-of-sight projected seismic wavefield and photospheric vector magnetic field at roughly 12 million pixels corresponding to the disk of the Sun. HMI, in comparison to MDI, takes high-quality, high-resolution observations of the visible full disk of the Sun, with fewer sources of systematical errors, allowing for seismic measurements to be taken closer to the limb.

1.4 Local Helioseismology

With improvements in observational techniques and the resolution of major problems of solar structure and rotation, the focus turned to the study of the 3-D Sun. One of the early contributions in this area was the introduction of ring-diagram analysis by Hill (1988), who envisaged measuring oscillation frequencies of relatively small-wavelength modes over spatially compact regions. For instance, oscillation frequencies measured over a limited region of the observed disk that had active regions would be qualitatively different from those in an equivalent non-magnetic 'quiet Sun' region, thereby shedding light on the sub-surface structure of magnetism. Rotation, flow systems, and other interior properties can also similarly be retrieved. While a robust technique, drawbacks of this method are that inferences are effectively averages over the unit tile of measurement, which can be quite large. To improve spatial resolution, these tiles have to be restricted in size, which in turn restricts the depth that can be imaged to near-surface layers. Therefore, in practice, only layers $r/R_\odot > 0.97$ can be studied using ring-diagrams.

To overcome some of these restrictions, Duvall et al (1993) introduced a powerful technique known as time-distance helioseismology, which is one of the most widely used local methods. Time-distance is an extension of prior seismic methods; while ring-diagram analysis and global-mode seismology use spatio-temporal power spectra, time-distance focuses on correlations of temporal series' recorded at disparate spatial points. See Figure 1.9 for observations of arrival times of different waves in the Sun. Because cross correlations involve point pairs, data handling complexity increases dramatically to an $O(N^2)$ problem, where N is the number of spatial pixels. Time-distance is termed as such because it relies on measuring wave travel times as a function of point-pair configurations. Like frequencies, the wave travel time is a natural measurement in seismology, connected directly to the wavespeed of the medium, which in turn is a proxy for temperature (sound speed), magnetic field, and flow speeds. In principle, one can derive 3-D maps of these properties using time-distance helioseismology.

Helioseismic holography (Lindsey and Braun 1997), another local technique, is based on optical holography methods. Holography, much as time-distance, uses temporal correlations between disparate spatial points, but differs in many other aspects. Surface wavefield measurements are separated into up- and downward propagating waves, and Green's functions of the Sun are used to propagate the waves in reverse time into the interior, from which properties such as magnetic fields, flows, and sound-speed perturbations may be retrieved.

1.4.1 Observational and theoretical Challenges

While these local methods continue to show great promise, two intrinsic limitations on imaging exist: stochastic wave excitation and the narrow modal bandwidth. Because waves are generated by the action of random granulation, the raw measured wavefield is described by a stochastic process (e.g., Woodard 1997). Consequently, one must take correlations of the wavefield to obtain meaningful seismic measurements, and because the temporal length of the data is finite, the correlations will possess realization noise. This stochastic noise limits the ability to precisely measure quantities such as modal frequencies and wave travel times, in turn diminishing the quality of seismic inference. A completely different problem is rooted in the very narrow temporal bandwidth of trapped resonant modes, extending from some 2 mHz to 5.5 mHz, less than a third of a decade in temporal frequency. This implies that there is a limited set of modes with which to image.

Further, imperfections in instrumental and other observational systematical errors such as image distortion, plate-scale deformation, etc. tend to strongly bias results. However, the most serious issues are unconnected with the instrument altogether. Line-of-sight projection ensures that different components of the (vector) wavefield velocity are recorded in different parts of the disk. For instance, the center of the disk will show predominantly radial velocities whereas closer to the limb, some combination of horizontal (lateral) and radial velocities come into play. More strangely, wave travel times, when recorded between point pairs at sufficiently large distances ($> 30°$ on the solar surface), suggest the existence of a horizontal outflow emanating from the disk center (i.e., in the plane of the solar disk). Evidently this is a measurement issue, since the Sun is rotating and the "disk center" is hypothetical, since it is just a point of observation. It is increasingly believed that granulation-related flows blue shift the wavefield (Baldner and Schou 2012), and because of the line-of-sight issue, there is a strong center-to-limb variation, thereby creating the semblance of an outflow. Inferences of meridional circulation are heavily reliant on being able to model these effects, since an aphysical (false) outflow signal from the disk center will pollute measurements. Annual orbital changes which result in variations in the angle between the ecliptic and the solar equatorial plane, the so-called B-0 angle, also contribute to systematic changes.

The high-frequency approximation, in which waves are treated as pencil thin rays, is widely invoked. However, there is evidence, both observational

(Duvall et al 2006) and theoretical (Gizon and Birch 2002), which suggest that 'finite-frequency' effects play an important role. Finite-frequency effects arise when wavelengths are comparable to the scale of the scatterer (such as a supergranule). These effects, while mathematically and computationally difficult to incorporate, must nevertheless be taken into account to ensure accuracy. A survey of helioseismic literature quickly reveals that finite-frequency theory is very rarely chosen. Ignoring these effects thus adds a systematic theoretical bias to seismic inferences.

This set of issues, the attendant controversies and inconsistencies have dented the reputation and success of local helioseismology. The launch of space-based observatories has made high-quality observations abundant and so there is now a serious need for the development of theory and novel techniques to enable accurate interpretation. A demonstrable, consistent, and uncontroversial success in local helioseismology can go a long way towards bolstering the case for the importance of this field of study. The motivation for all forms of seismology, local and global, has always been strong, namely that of understanding stellar physics, the dynamo, convection, etc. Global helioseismology has delivered on its promise but local methods have yet to achieve their vaunted goals of being able to reliably infer meridional circulation, convection, and magnetic fields in the interior.

1.5 The Forward Problem

The forward method is the connection between the model of the Sun and the wavefield; it is the means of calculating the effect of models on the seismic wavefield. Thus one conceivable way to infer solar interior structure is to compute seismic signatures associated with a broad range of possible forms of the feature of interest (such as all types of meridional circulations, etc.) and compare with observations. Thus the wavefield is computed in the presence of various sizes, magnitudes, and (perhaps) types of perturbations; relevant seismic measurements obtained thereof are compared to those obtained observationally. These calculations are carried out with the outlook that a close approximation to observations is the reason to believe that the structure of the feature is, to some extent, representative of reality. Further, by exploring the *model* space, it is possible to comment on the degree of uniqueness of final inference. This would correspond to the variance in model space, i.e., aspects of models that cannot be distinguished given the uncertainty in and/or the insensitivity of seismic measurements. This is the *forward* approach, in that it goes from model space to data space. A variety of techniques may be applied to execute the forward calculation, including ray theory, Rytov approximation, Born approximation, and full-wave simulations (see Gizon and Birch 2005; for a review). As observations have become increasingly sophisticated, the need for refined forward modeling has become apparent. Among these approaches, full-wave simulations are the most complete and capable of accounting for all relevant effects. But a number of outstanding issues remain, one at the very heart of the forward approach: the governing equation.

1.5.1 The Wave Equation

Velocities of individual modes in the Sun, on the order a few cms^{-1}, are very small in comparison to the local sound speed ($> 7kms^{-1}$). The linear wave equation therefore serves, one would imagine, as a perfectly valid starting point. It is however remarkable that it is not only a good starting point but a general and very accurate approximation. The discrepancy between observationally derived frequencies and those predicted by standard solar models, i.e., the normal modes of a sphere whose stratification is given by a standard solar model, is as small as 0.1% for a variety of modes. Despite the recognition that the Sun is anything but standard and that waves propagate through a convecting medium, early successes have led to the widespread adoption of the simple wave equation. This simple wave equation is incomplete. Equally widespread is the recognition that convection both phase shifts waves in a systematic manner and causes the overall decoherence of resonant modes due to scattering. The current practice is to apply empirical corrections but a theory of frequency shifting and wave damping is needed.

1.5.2 Solving the Wave Equation

1.5.3 Convective Instability

The outer third of the Sun (by radius) is in a constant state of the convection and solar structure models suggest that near-surface layers exhibit vigorous convection. The onset of convection occurs due to an instability in stratification, most popularly expressed in terms of a negative Brunt-Väisälä frequency squared ($N^2 < 0$). Small-amplitude perturbations around a convectively unstable stratified medium will undergo exponential growth (the full non-linear equations are stable since the non-linear terms would act to stabilize the system). A typical approach then is to take standard models of the Sun, such as model S (Christensen-Dalsgaard et al 1996), and alter the stratification of the near-surface layers to ensure convective stability while simultaneously ensuring hydrostatic balance and an appropriate cutoff frequency (Hanasoge et al 2006; Parchevsky and Kosovichev 2007; Hanasoge et al 2008). Waves at temporal frequencies below the cutoff are reflected by the sharp density gradient (small density scale height) in surface layers. Thus, to ensure that the solution of the equation contains the same set of modes as the Sun, it is important that the cutoff frequency be reproduced.

Other approaches involve formulating the governing wave equation in a way such that N^2 appears explicitly and forcibly setting it to zero, i.e., $N^2 = 0$. This method faces the problem that the cutoff frequency is reduced from its nominal value of ~ 5.5 mHz for the Sun to about 3.5 mHz (Hartlep et al 2008a).

The likely best technique, as yet unimplemented, is to drop the initial-value-based temporal evolution altogether, instead transforming the governing equation to temporal Fourier domain. The equation is then solved as a boundary-value problem at frequencies where the wave equation is stable (convective instabilities occur at frequencies less than 2 mHz). Further, because the wave equation we deal with here is linear, the frequency-domain representations of solutions to the equation at two different frequencies are independent of each other. As a consequence, the frequency domain is an attractive regime in which to solve the wave equation. However, this formulation poses unsolved numerical challenges such as the design of stable and rapidly convergent schemes.

1.5.4 Magneto-hydrodynamic Singularities and Waves

To retrieve the structure of sub-surface magnetic fields from seismology, it is essential to predict seismic signatures associated with these fields, and prediction is reliant on numerical simulations of wave propagation through magnetic fields. Seismically, the structure of magnetic media is sensed by magneto-acoustic slow and fast modes and Alfvén waves, each of which propagates at a different wavespeed. One of the pathological properties of the magneto-hydrodynamic (MHD) model is that Alfvén waves cannot exist in the absence of magnetic fields. In other words, a medium where the magnetic flux is tightly concentrated in a bundle, e.g., sunspots or magnetic elements in the Sun, Alfvén waves, which are essentially shear waves, can only exist and propagate within this flux region. This is problematic since at the edges of the flux concentration, where the field weakens arbitrarily, the wavelength of the Alfvén wave ($\lambda = c_A/\nu$, where λ is the wavelength, c_A is the Alfvén speed, and ν is frequency) becomes vanishingly small and the equation governing these waves becomes singular. The Alfvén speed is positive *semi*definite, thereby creating difficulties for numerical simulations.

The density of the solar plasma falls exponentially rapidly with height above the photosphere (scale height of around 200 km). Commensurately, the hydrostatic gas pressure falls very quickly. The divergence-free nature of magnetic field constrains magnetic structures in a manner that the (magnetic) Lorentz to hydrodynamic force ratio becomes extremely large. The Alfvén wavespeed, given by $|\mathbf{B}|/\sqrt{4\pi\rho}$, where $|\mathbf{B}|$ is the magnitude of the magnetic field and ρ is density, correspondingly takes on extremely large values (hundreds or even thousands of km/s). This poses grave numerical difficulties, and a variety of unsatisfactory approximations to mitigate the problem have been proposed (Cameron et al 2008; Hanasoge 2008; Rempel et al 2009; Moradi and Cally 2014).

1.5.5 Wave Damping

Waves are damped in a complicated manner in the Sun, possibly by scattering due to convection, radiative losses, and/or non-linear effects. Linear simulations cannot hope to capture all these effects. Consequently, the Sun is effectively treated as a damped harmonic oscillator, where physical damping is replaced by some phenomenological frequency-dependent attenuation term. This term is charged with the task of incorporating observed wave damping rates, which are sensitive functions of frequency. High-fidelity simulations require that the full damping characteristics be modeled, but efforts to achieve the complicated dependence of attenuation with frequency have not succeeded for initial-value problems. Again, one likely solution appears to be that of the boundary-value approach, and because the equation is written in Fourier domain, the frequency-dependent damping term becomes multiplicative (a convolution in temporal domain) and can therefore be incorporated.

1.6 Chapters in the book

Contemporary helioseismology deals with the study of the 3-D structure of the Sun, the imaging of large-scale fluid circulations and magnetic fields in the interior. Solving these problems requires the application of a variety of unconventional and novel mathematical and computational techniques.

The structural complexity of the 3-D Sun essentially rules out purely analytical lines of investigation. Numerical techniques play a central role in helioseismology and the development of methods is critical. The sound speed of the Sun varies by almost two orders in magnitude from the center to the surface, and the density by

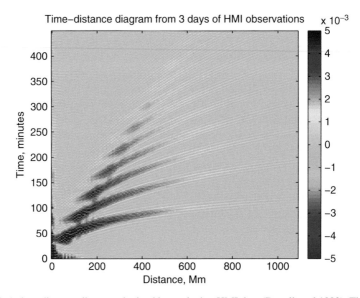

Fig. 1.9 A time-distance diagram obtained by analyzing HMI data (Duvall et al 1993). The ridges seen in the figure correspond to the time (on the vertical axis) taken by wave packets to travel the distance (in degrees) shown on the horizontal axis. The wave packet structure is contained in the ridges. The first bounce represents the shortest acoustic path between two points at the surface, whereas the second bounce records waves that proceed from the first point, bounce once at the surface and then arrive at the second point, and so on.

twelve orders. The design of a precise and stable numerical method and testing it to ensure accuracy of the solution is a task of critical importance. It is also important to incorporate high-fidelity absorbing side and bottom boundaries for local simulations (of course the whole Sun has only one outer radial boundary). Numerical methods to solve the wave equation and the attendant difficulties that are encountered are discussed in Chapter 2.

Post-equation formulation and post-numerical development, the question centers around how best to use the forward solver. For instance, given a set of seismic measurements at the surface, the determination of internal properties of the Sun could proceed by computing the seismic signatures associated with a suite of different structural models. However, this probabilistic approach is expensive and therefore generally infeasible. A more relevant methodology relies on defining a misfit between observation and prediction and computing the gradient of the misfit function with respect to the model parameters. Gradient-based optimization is widely utilized in terrestrial seismology and airfoil design, and is very powerful, especially when starting from a good starting guess. See Figure 1.10 for a schematic of the steps involved. The adjoint method, a technique used in control theory, allows for the rapid computation of gradients and will be the subject of Chapter 3.

Developing stable and accurate numerical methods to simulate the helioseismic wavefield and applying the adjoint method to compute the gradient of the misfit function all lead up to the goal of helioseismology: inferring properties of the interior. Chapter 4 deals with the solution of large inverse problems, specifically focused on full waveform inversion. With increasing sophistication in methodology, we can use more of the seismic measurement, the eventual goal being that of fitting the *full* waveform, thereby using all available seismic information (and hence the name).

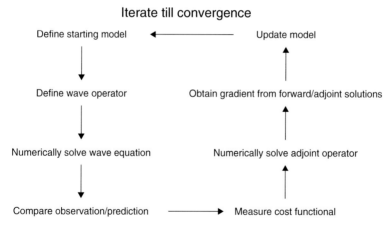

Fig. 1.10 How to perform an inversion using the adjoint method. The procedure begins with the definition of a suitable background (structure) model and the wave operator. Solutions to the operator give the predicted wavefield, which may then be compared with observations in order to characterize the cost functional. Output from the forward solver is then fed into the adjoint calculation. The forward and adjoint solutions are then temporally convolved to obtain the gradient of the misfit function with respect to all the model parameters. The gradient is then used to update the background model and the process is repeated.

Chapter 2
Wave Equation Solver[†]

The fast computation of solutions to the helioseismic wave propagation problem can help propel us closer to the goal of accurately inverting for the sub-surface structure and dynamics of the Sun. It also improves our understanding of wave physics, assists us in interpreting the helioseismic measurement response to the presence of anomalies in the interior and can play a central role in obtaining solutions of the inverse problem (adjoint method; Hanasoge et al 2011; also chapter 4 of this monograph). Towards this end, numerous codes and methodologies exist (Hanasoge et al 2006; Shelyag et al 2006; Khomenko and Collados 2006; Hanasoge et al 2007; Cameron et al 2007; Parchevsky and Kosovichev 2007; Hanasoge 2008; Hartlep et al 2008a; Felipe et al 2010). Several advances have been made through the use of these simulations: in addressing the imaging of hydrodynamic phenomena within the interior, see, e.g., Hanasoge and Larson (2008); Hartlep et al (2008b); Hanasoge and Duvall (2009); Zhao et al (2009); Hanasoge et al (2010a); wave phenomena associated with magnetic regions have been studied by, e.g., Khomenko and Collados (2006); Hanasoge et al (2008); Cameron et al (2008); Hanasoge (2008); Parchevsky et al (2008); Moradi et al (2009); Khomenko and Collados (2009); Hanasoge (2009); Birch et al (2009); Shelyag et al (2009); Cameron et al (2010); Felipe et al (2014).

[†] The content of this chapter is taken from Hanasoge et al (2006), Hanasoge and Duvall (2007), Hanasoge (2008), Hanasoge et al (2008), Hanasoge et al (2010b). The spherical solver is available on request and the Cartesian code can be downloaded from http://www.tifr.res.in/~hanasoge/sparc.html

2.1 Equations

We assume that the background model satisfies

$$-\nabla p_0 - \rho_0 g_0 \mathbf{e}_z + \frac{1}{4\pi}\mathbf{j}_0 \times \mathbf{B}_0 = 0, \tag{2.1}$$

where all quantities with subscript '0' are time-stationary and considered to be part of the background; p is the pressure, ρ density, g_0 gravity, $\mathbf{B}$ magnetic field, $\mathbf{j} = \nabla \times \mathbf{B}$ current density. Rotation is neglected and hydrodynamic forces due to flows are for most practical purposes very small and do not contribute to maintaining the background equilibrium. Consequently, the wave equations are "God-given" at some level since terms such as damping and flow advection do not follow directly from the equilibrium equation (2.1). The assumption of linearity is justifiable since acoustic wave velocity amplitudes (order of a few cm/s) are much smaller than the background sound speed (order of several km/s) within the bulk of the computational domain. We may pose the linearized equations (solved in the bulk) in terms of vector displacement ($\boldsymbol{\xi}$) or velocity ($\mathbf{v}$). Since these are written for weak flows (note this requires $\|\mathbf{v}_0\| \ll \omega L_0$, where L_0 is a characteristic length scale of the flow and ω is the wave frequency), we neglect second-order terms in $\mathbf{v}_0$. Respectively, the equations for displacement and velocity are:

$$(\partial_t^2 \boldsymbol{\xi} + \Gamma \partial_t \boldsymbol{\xi}) = -2\mathbf{v}_0 \cdot \nabla \partial_t \boldsymbol{\xi} - \frac{1}{\rho_0}\nabla p - \frac{\rho g_0}{\rho_0}\mathbf{e}_r$$

$$+ \frac{1}{4\pi\rho_0}(\mathbf{j}_0 \times \mathbf{B} + \mathbf{j} \times \mathbf{B}_0) + \mathbf{S} \tag{2.2}$$

$$p = -c_0^2 \rho_0 \nabla \cdot \boldsymbol{\xi} - \boldsymbol{\xi} \cdot \nabla p_0, \tag{2.3}$$

$$\rho = -\nabla \cdot (\rho_0 \boldsymbol{\xi}), \tag{2.4}$$

$$\mathbf{B} = \nabla \times (\boldsymbol{\xi} \times \mathbf{B}_0), \quad \nabla \cdot \mathbf{B} = 0, \tag{2.5}$$

$$\mathbf{j} = \nabla \times \mathbf{B}, \quad \mathbf{j}_0 = \nabla \times \mathbf{B}_0, \tag{2.6}$$

where $\mathbf{S}$ is the forcing function (of space and time), and Γ is a pre-specified damping function. Boundary treatments may require the introduction of additional terms/equations. Note that in Cartesian geometry, $\mathbf{e}_r$ is replaced by $\mathbf{e}_z$. Equation (2.3) is a statement of adiabatic wave propagation and is justified on the basis of the long viscous and heat transfer timescales in comparison to the acoustic timescales over much of the solar interior.

2.2 Convective Instability

Computationally, the properties of the Sun are well behaved and easy to model up to about $r = 0.98R_\odot$. The near-surface layers however, introduce the multiple difficulties of rapidly dropping density height scales, increasingly unstable stratification, the presence of ionization zones, complexity in the equation of state, and possibly non-linearities into the wave propagation physics. Added to these issues is the fact that acoustic waves spend most of their time in the near-surface layers because the sound speed is the smallest here.

The Brünt-Väisälä frequency indicates whether or not a medium is unstable to convection. It is given by (see Christensen-Dalsgaard et al 1996; chap. 3)

$$N^2 = g\left(\frac{1}{\Gamma_1}\frac{\partial \ln p}{\partial r} - \frac{\partial \ln \rho}{\partial r}\right), \qquad (2.7)$$

where g is gravity, N is the Brünt-Väisälä frequency, and Γ_1 (known as the first adiabatic exponent) is defined as

$$\Gamma_1 = \left(\frac{\partial \ln p}{\partial \ln \rho}\right)_{ad}, \qquad (2.8)$$

where the derivative is evaluated along an adiabatic process curve (as denoted by the subscript 'ad'). The solar convection zone extends all the way from roughly $0.7R_\odot$ to the surface. For purposes of discussion, we shall divide the convection zone into two regions, $0.70R_\odot < r < 0.996R_\odot$ where timescales of convective growth are considerably larger than acoustic timescales (5 minutes) and $0.996R_\odot < r < 1.0003R_\odot$ where the convective growth rate and acoustic timescales are comparable. Consider the inner region with slowly growing instabilities first. We are dealing with a linear system and at first sight, it seems odd that although we restrict acoustic excitation to the bandwidth $2.2 - 5.2$ mHz, we still see instabilities at much lower frequencies. The reasons for this are the finiteness of the excitation time series, which results in the broadening of the frequency response, and numerical round-off errors, which act as broadband sources. This exponential growth significantly boosts power at low frequencies within a few hours into the calculation, making multi-day simulations impossible. In order to perform simulations, we alter the background model in the manner described in, e.g., Hanasoge et al (2008) such that the cutoff frequency is unchanged while ensuring $N^2 \geq 0$.

The outer convective envelope introduces difficulties which must be treated with greater care. As can be seen in Figure 2.1, the instability timescales very close to the surface coincide with the center of the acoustic bandwidth. Since our interest lies in capturing the interaction of the acoustics with the background dynamics and not in the direct computation of the convection, we must devise a means to remove this instability without affecting the acoustics. One way to accomplish this is to alter the Brünt-Väisälä frequencies. A crucial requirement is that the acoustic impedance of the surface layers not be changed by much, since all the acoustic reflection occurs in and around these layers. The force-balance term, ρg, in the pressure equation (2.3) introduces convective instabilities into the eventual oscillation equation.

In general, changing the structure of the near-surface layers is a non-ideal solution since changing the background shifts the eigenfrequencies and eigenfunctions of the resonant modes. Mathematically speaking, since the wave operator (without damping and flows) is Hermitian (e.g., Lynden-Bell and Ostriker 1967), all eigenvalues (ω^2) for this operator are real, which in itself is a useful piece of information, since it tells us that all the poles lie either on the real or imaginary axes and nowhere else (note that ω^2 is real which means ω is either purely real or purely imaginary). The question then is how do we shift the unstable poles on the imaginary axis onto the stable complex half-plane?

Thus there has been a gradual interest in adopting methods of active instability control (i.e., by applying some sort of frequency filter), but a clear idea of how

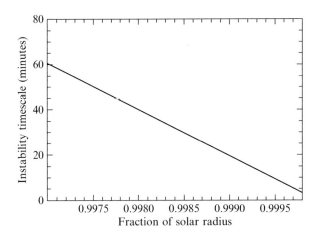

Fig. 2.1 Convective instability timescales $2\pi / \sqrt{|N^2|}$ as a function of the non-dimensional radius. At $r = 0.9993R_{\odot}$ and greater, the linear instability starts directly interacting with the acoustics and corrupting the signal. The instability arises as a direct consequence of the super-adiabaticity of the background model, and since we are not modeling the non-linear physics of convection, it is crucial that we prevent this linear instability from affecting the acoustic signal.

to successfully do so remains elusive. Alternately, the equations can be solved in frequency-wavenumber space, thereby excluding unstable gravity modes, but designing a boundary-value solver would then become necessary.

2.3 Solver in spherical geometry

When simulating waves whose wavelengths are comparable to the solar radius, it is important to take into consideration spherical geometry. Placing a latitude-longitude-radius grid to resolve a 3-sphere is the simplest and best established way to pose the numerical problem. However this grid contains coordinate singularities at the poles and at the centre. While the former may be addressed by using spherical harmonics, the latter (central singularity) is more difficult to deal with. Consequently, we describe the solution of the three-dimensional equations of fluid motion in a spherical shell encompassing $0.2 - 1.0004R_{\odot}$, linearized around a convectively stable form of the spherically symmetric background state described by Model S of the Sun (Christensen-Dalsgaard et al 1996; Hanasoge et al 2008).

It is believed that wave excitation in the sun occurs in an extremely narrow spherical envelope ($\sim$ 100 km thick) bounded by the surface (e.g., Nordlund et al 2009), and we assume therefore, $S(r,\theta,\phi,t) = \tilde{S}(\theta,\phi,t)\delta(r - r_{ex})$, where $r_{ex} = 0.9997R_{\odot}$ was chosen to be the radial location of the source. $\tilde{S}$ is a spatially broadband random function for all but the largest horizontal wavenumbers, which are not included so as to avoid issues of spatial aliasing. The solar acoustic power spectrum possesses maximum power in the range $2 - 5.5$ mHz with a peak in power around 3.2 mHz. In order to mimic this excitation behaviour, we generate a Gaussian distributed power spectrum with a mean of 3.2 mHz and a standard deviation of 1 mHz in frequency space, which we then Fourier transform to produce a time series with the appropriate source spectrum. Because timescales of acoustic propagation are generally much smaller than the timescale over which large-scale flows or features (of interest to us) change, we assume that the background state is stationary.

It may be useful to perform simulations over time periods that exceed the time at which the acoustic energy reaches a statistical steady state. The other requirement for the temporal length of the simulation is that the frequency resolution be sufficient for the application at hand. The velocity time series, extracted at the surface, is projected onto a line of sight and used as artificial Doppler velocity data.

2.4 Numerical method

The procedure we employ is pseudo-spectral; we use a spherical-harmonic representation of the spherical surface, sixth-order compact finite differences in the radial direction (see Lele 1992) and a fourth-order, five-stage Low Dissipation and Dispersion Runge-Kutta (LDDRK) time-stepping scheme (see Hu et al 1996). Latitudes are Gaussian collocation points and longitudes are equidistant. The choice of the radial grid is discussed in section 2.4.1.

The code runs in parallel, written according to the Message Passing Interface 1.2 standard, with latitudes distributed across processors, and all longitudes and radial points corresponding to each latitude located in-processor. Spherical-harmonic transforms are computed in two steps: longitudinal Fast Fourier Transforms (FFTs) at each latitude and radius followed by Legendre transforms for each Fourier coefficient and radius. FFTs are performed using the Guru routines provided in the freely available package FFTW and Legendre transforms using matrix-matrix multiplication techniques implemented in Basic Linear Algebra Subroutines (BLAS) packages. The FFTs can be performed locally since all longitudes are in-processor. In order to perform the Legendre transform, we transpose each array so that all latitudes for a given radial and longitudinal point are located in-processor. The associated Legendre polynomials $P_{\ell m}$, where ℓ and m are the spherical-harmonic degree and order, respectively, are divided into a series of matrices corresponding to different m's, each of which is further divided into two matrices according to whether $(\ell - m)$ is even. This is done to exploit symmetries in the associated Legendre polynomials, which speeds up the transform by a significant amount.

The associated Legendre polynomials are computed according to a stable four-term recurrence algorithm given by Belousov (1962), computed once at the start of the calculation. Each transform is a computation of order $O(n_{\text{lon}}^2 \cdot n_{\text{lat}} \cdot n_{\text{rad}} \cdot \log(n_{\text{lon}}))$, where n_{lon} is the number of longitudinal grid points, n_{lat} the number of latitudinal grid points, and n_{rad} the number of radial grid points. To prevent aliasing, we apply the two-thirds rule (Orszag 1971) which sets the lower bound on the number of latitudes at $3\ell_{\text{max}}/2$ where ℓ_{max} is the spherical-harmonic bandwidth. To ensure equal resolution on the spherical surface, we set $n_{\text{lon}} = 2n_{\text{lat}}$. Recasting the minimum operation count in terms of ℓ_{max}, we arrive at an expensive $O(n_{\text{rad}} \cdot \ell_{\text{max}}^3)$; it is therefore important to minimize the number of times spherical-harmonic transforms are performed. Every timestep requires the computation of a curl, divergence, and four gradients, each of which involves a computational equivalent of a forward-inverse transform pair. BLAS is known to operate near the peak performance of the processor, so these computations are generally very efficient, when they are performed in-processor.

We place radiating boundary conditions (Thompson 1990) at both radial boundaries of the computational domain. However, this particular boundary condition is most effective at absorbing waves that are of normal incidence but reflects a significant percentage of all other waves. To mitigate this effect, we introduce an absorbent buffer zone (for example, see Colonius 2004; Lui 2003), placed in the evanescent region, that damps waves out substantially before they reach the boundary. This is one of the purposes that the term $\Gamma(r)$ in equation (2.2) fulfills (see section 2.4.2).

2.4.1 Choice of radial grid

Although background properties such as pressure and density depend strongly on radius, sound speed is the most important determining factor in setting the physics of wave propagation over most of the interior. Consider therefore a wave propagating at the speed of sound in the radial direction according to the simple advection equation

$$\partial_t u + c(r)\partial_r u = 0. \tag{2.9}$$

It makes sense to choose a grid stretching function

$$\tau(r) = \int_r \frac{dr'}{c(r')}, \tag{2.10}$$

that transforms equation (2.9) to

$$\partial_t u + \frac{\partial u}{\partial \tau} = 0, \tag{2.11}$$

a form that is much easier to handle. The relation between two adjacent grid points
then is

$$\int_{r_i}^{r_{i+1}} \frac{dr}{c} = \delta,$$ (2.12)

$$\delta = \frac{1}{n_{rad} - 1} \int_{r_{in}}^{r_{out}} \frac{dr}{c},$$ (2.13)

where r_{in}, r_{out} are the inner and outer radii, respectively, and n_{rad} is the number of
radial grid points including the boundaries. Since sound speed is a monotonically
decreasing function of radius, the radial grid spacing is larger at depth. Also important to note is that gradients of background quantities become smaller with depth
and it makes sense that the grid is coarser. Figure 2.2 displays grid spacing as a
function of radius.

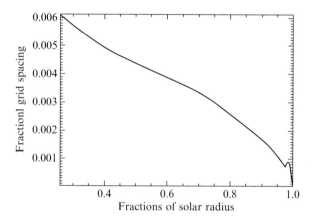

Fig. 2.2 Fractional grid spacing as a function of radius for $n_{rad} = 300$. Plotted is $\Delta r/R_{\odot}$, where
Δr is the local grid spacing, as a function of the fractional radius, $r/R_{\odot}$. For $r \leq 0.98R_{\odot}$, the grid
spacing is chosen to maintain the constancy of the travel time of an acoustic wave between adjacent
grid points. To account for the rapidly decreasing density and pressure scale heights, the radial grid
points from $0.985R_{\odot}$ to the upper boundary are equally spaced in logarithmic pressure. Third-order
splines are used to vary grid spacing between 0.98 and $0.985R_{\odot}$ as smoothly as possible.

2.4.2 Buffer layer

It was mentioned earlier that the transmitting boundary conditions employed in this
calculation reflect a large percentage of waves that impact it at significant angles
(as opposed to purely radially propagating waves). This can impact both the stability and accuracy of the simulation. In order to deal with this problem, we insert a
buffer layer adjoining the boundary meant to damp waves out significantly before

they reach the boundary. This ensures that even if these waves are reflected at the boundary, they will have to propagate through the buffer layer again to reach the computational region of interest. This layer serves to diminish the amplitudes of these aphysical waves to insignificance. We place a buffer layer at each end of the computational domain to prevent unwanted reflection.

2.4.3 Spectral blocking and radial dealiasing

Spectral blocking is an aliasing phenomenon that commonly occurs in non-linear calculations, wherein the lack of resolution results in the accumulation of energy near the Nyquist frequency (Hanasoge and Duvall 2007). It poses a serious numerical challenge, since the energy at the Nyquist grows with each timestep, leaving the computation unstable and eventually inaccurate. We discuss its appearance in our *linear* calculations and how we deal with this issue. Standard Fourier transforms are defined on grids where the travel time for waves between adjacent grid points is a constant over the grid. In the solar case, the sound speed is a strong function of radius and consequently, it makes little sense to speak of a Fourier transform on a uniformly spaced radial grid. The Fourier transform in this situation is meaningful on a grid stretched such that the travel time between adjacent grid points is constant over the grid. The rest of the discussion in this section follows as a consequence of this grid stretching and the consequent interpretation of the Fourier transform on this grid.

In order to mimic wave excitation in the Sun, we place sources that are highly localized in the radial direction, resulting in the generation of waves at broad range of radial orders. The resolution in the radial direction is restricted by the finiteness of computational resources at our disposal and the scientific interest in investigating these high radial orders. For the applications that we are interested in, both these criteria indicate that these high radial orders are best done away with. Associated with the inability of the radial grid to capture modes containing rapid variations is the phenomenon of aliasing which causes waves beyond the resolvable limit of the grid to fold back across the Nyquist onto the resolvable waves near the Nyquist. This by itself is not a serious problem since we are only interested in a small number of ridges that are situated well away from the radial Nyquist. Typically, aliasing in linear problems is relatively harmless and usually only results in a slight increase in power near the Nyquist.

Fourier transforms in the radial direction display spectral blocking, an effect that occurs in numerical solutions of non-linear equations, commonly seen in simulations of turbulence and other non-linear phenomena. It is seen in our computations because of the highly non-constant terms (in the solar case) of the wave equation, density, pressure, and sound speed, that premultiply the linear fluctuation terms, such as the first term on the right-hand side of equation (2.4). These non-constant terms act as conveyor belts across the radial spectrum, transferring energy between disparate wavenumbers, and eventually cause this an energy build-up at the Nyquist.

This energy accumulation can destabilize the calculation and diminish its accuracy. In order to dealias the variables, we apply a 11-point dealiasing filter (Vichnevetsky and Bowles 1982) described in Hanasoge and Duvall (2007). Because of the high order of the filter, the portion of radial spectrum of interest is left largely unaffected. For further details on validation and verification, the reader may refer to Hanasoge et al (2006).

2.5 Solver in Cartesian geometry

A significant thrust of (local) helioseismology is the study of features in the Sun whose spatial scales are small in comparison to the solar radius (~ 700 Mm), such as sunspots (~ 40 Mm), supergranules (~ 30 Mm), and granules (~ 2 Mm). In these local regions, spherical geometry may be ignored and the problem may be reduced to simulating wave propagation in Cartesian computational domains. Because magnetic fields strongly influence wave propagation in the near-surface layers (e.g., Moradi et al 2009), the code needs to solve the linearized magneto-hydrodynamic (MHD) wave equation. The rest of this chapter will focus on the construction of a numerical method and attendant issues.

SPARC is a publicly available code for helioseismology that solves the linearized MHD wave equation in Cartesian geometry. SPARC is validated and documented, maintains parallel efficiency, makes use of modern numerical methods, and incorporates a stable and accurate absorbing boundary formulation based on the perfectly matched layer for wave absorption in stratified, MHD media (Hanasoge et al 2010b). It incorporates realistic wave damping (the prescription of Schunker et al 2011) and is able to stably simulate wave propagation through numerous models of sunspots. Differing forms of Lorentz-force limiters are implemented in the code, e.g., Moradi et al (2009); Hanasoge et al (2010b); Cameron et al (2010); Moradi and Cally (2014). We note that there are other codes, some of which are also publicly available, such as by, e.g., Cameron et al (2008), Khomenko and Collados (2006) that incorporate a variety of numerical methods. In this section however, we shall focus on SPARC.

The primary focus of this chapter is how numerical issues are dealt with in SPARC; further detail on the development, validation, and verification of the numerical methods may be found in Hanasoge et al (2006), Hanasoge et al (2007), Hanasoge and Duvall (2007), Hanasoge (2007), Hanasoge (2008), Hanasoge et al (2008), Hanasoge et al (2010b). Currently it is used by a number of helioseismologists; the code itself is available on the following website: http://www.tifr.res.in/~hanasoge/sparc.html.

2.6 Numerical Implementation

2.6.1 Stratification

Depending on where the vertical boundaries of the model are fixed, there may be as many as 10–15 scale heights in density between the deepest and highest grid points. An immediate question to ask is if this strong level of stratification introduces a stiffness to the problem and whether the timestep is rendered small as a consequence. Such an effect would make the computational burden excessive and simulations may then not be a reasonable way to proceed. However, it may be demonstrated (e.g., Christensen–Dalsgaard 2003) that with the exception of a small near-surface region, acoustic waves are in fact largely insensitive to the degree of density stratification, reacting primarily to the sound-speed distribution, which in comparison varies much more slowly with radius. The consequence of this analysis is that numerical simulations of wave propagation in solar-like media are tractable and computations can proceed in a finite and reasonable amount of time. The vertical grid is set according to the recipe described in Section 2.4.1.

2.6.2 Spatio-temporal schemes and parallelism

We make use of modern numerical methods in the computation of spatial derivatives and to evolve the system forward in time. This is necessary in order that the computation be accurate and efficient. A sixth-order compact (implicit) finite difference method with fifth-order accurate boundary conditions (Lele 1992; Hurlburt and Rucklidge 2000) is applied to compute vertical derivatives and FFTs (Cooley and Tukey 1965) are used to estimate horizontal derivatives. Options to use alternate high-order methods (ninth-order explicit differences; Berland et al 2006) to compute spatial derivatives have also been included. The solution is temporally evolved using a second-order optimized Runge-Kutta integrator (Hu et al 1996). It incorporates realistic wave damping (the prescription of Schunker et al 2011) and is able to stably simulate wave propagation through numerous models of sunspots, with the application of differing forms of Lorentz-force limiters, e.g., Moradi et al (2009); Rempel et al (2009); Hanasoge et al (2010b); Cameron et al (2010); Moradi and Cally (2014).

The sizes of the problems of study can be large (e.g., $512 \times 512 \times 300$ grid points), requiring distributed computing. The parallelism is concordant with the Message Passing Interface standards v1.2 and higher. Each variable is an array of dimensions (n_x, n_y, n_z), where the three components are the number of grid points in directions (x, y, z), respectively. The algorithm involves dividing the arrays into chunks among processors as equitably as possible. Denoting the number of processors by n_P, the storage pattern is determined such that each processor contains approximately $(n_x, \lfloor n_y/n_P \rfloor, n_z)$ sized chunks of the original array (the symbol $\lfloor x \rfloor$ denotes the integer floor of positive real number x). Thus the parallelism is only in one direction; this makes the book-keeping much easier when interchanging information between processors. It is also viable when the problem is not too computationally expensive. Thus for very large sizes, i.e. when $n_x > 700$ or so, a parallel algorithm that involves domain decomposition along multiple directions may be more efficient. In any case, the derivatives in the (x, z) directions may be computed in-processor while for the y gradient, the arrays must be reconfigured such that all the y points for a given (x, z) are in the same processor. Using a series of nonblocking sends and receives, we are able to mask communication by computation.

Because the differential equations contain products of temporally evolving terms with background properties that vary dramatically in the vertical direction such as density and sound speed, the solution tends to be strongly aliased in z (see section 2.4.3 and Hanasoge and Duvall 2007). Further, when strong, spatially localized anomalies such as flux tubes are present, aliasing also occurs in the horizontal directions. In order to avoid spectral blocking, we invoke Orszag's 2/3 dealiasing rule (Orszag 1971) and apply filters in the horizontal and vertical directions (described in Hanasoge and Duvall 2007; and in Section 2.4.3).

2.6.3 MHS models and large Lorentz forces

A significant difficulty that has yet to be broadly dealt with is that of constructing an MHS model. Currently, a widely used model is that of self-similar magnetic tubes (Schlüter and Temesváry 1958). Unfortunately, it is not accurate in the atmosphere because the Lorentz forces remain large while the hydrostatic force falls exponentially (e.g., see, Moradi et al 2010), leading to completely evacuated flux-tube interiors (or even to negative pressures). Presumably, sunspot models must be constructed with the sub-photospheric layers in force balance and the atmospheric region force-free (as is likely in the Sun). As yet this remains an unsolved problem although there have been some efforts in this regard (e.g., Khomenko and Collados 2008). In any case, the code can currently handle arbitrary MHS models as inputs.

One of the biggest problems in simulating wave propagation in the magnetic Sun is the excessively large magnitude that the Alfvén speed ($c_A = ||\mathbf{B}_0||/\sqrt{4\pi\rho_0}$) attains in the atmosphere. This results in an extremely stiff problem with the timestep (Δt) being highly constrained by the Courant condition in the upper layers (i.e., $\Delta t \sim \Delta z/c_A$, where Δz is the vertical grid spacing). Consequently, in order to be able to compute in a finite amount of time, it has been the classical approach to introduce a Lorentz-force fudge factor that limits it when the ratio between Lorentz and hydrodynamic forces becomes too large (or in other words, c_A/c_0, where c_0

is the sound speed, becomes very large). The impact of this approximation on the wavefield is gradually being understood and while some authors have commented on the sensitivity of the solution to the specific choice of this limiter (e.g., Hanasoge 2008), the first serious piece of work on appreciating the Lorentz-force limiter was by Moradi and Cally (2014). Some choices for these limiters have been discussed before (Cameron et al 2008; Rempel et al 2009) and these prefix the Lorentz-force terms in the momentum equations. In a hand-waving sense, these reduce the Lorentz force while not affecting the divergence of the magnetic field fluctuation terms. However, this method results in a model that is not seismically reciprocal (Hanasoge et al 2011; and Appendix A.1), an important requirement in the formal interpretation of helioseismic measurements. Seismic reciprocity is a statement on the interchangeability of the source and receiver, i.e. that the wavefield measured at the receiver should be identical if the source-receiver points are swapped.

In SPARC, we implement a different form of the limiter: we directly alter the background magnetic field so that locally, the Alfvén speed never exceeds a certain predetermined value. In other words, at each height, we saturate the absolute value of the background magnetic field; this post-facto altered field typically does not obey $\nabla \cdot \mathbf{B}_0 \neq 0$. However, as seen from the oscillation equations, this does not affect the divergence-free condition on the fluctuations (equations [2.2] through [2.5]). In the framework of this form of the limiter, we find that the z-derivative terms in $\mathbf{j}_0$, i.e., $\partial_z B_{0x}, \partial_z B_{0y}$ become very large in the atmospheric layers, and cause instabilities near the upper boundary. Consequently, we drop these terms, arguing that in the Sun, it is likely that the variation of the background horizontal fields with height, especially in the atmosphere, is very small. This renders the calculation stable.

2.6.4 Vertical boundary conditions

A fairly important part of the numerical treatment concerns the boundaries. Since the Sun lacks a clear upper boundary, and because high-frequency and Alfvén waves likely leak into the corona and are dissipated there, the typical approach is to place absorption conditions at both vertical boundaries (the lower boundary too because it is a purely computational construct and we do not want waves to reflect off it). Lining the boundaries with sponges is a commonly implemented technique (and is a robust option in SPARC) and found to be a stable and an accurate way of absorbing waves (e.g., Hanasoge 2007; also see section 2.4.2). One drawback with using a sponge is that it is computationally expensive, requiring approximately 20 -30 grid points adjacent to each boundary. In contrast, a much more efficient (in terms of absorption and computation) method is that of the perfectly matched layer. With the development of the *stable* un-split convolutional perfectly matched layer (Hanasoge et al 2010b), now incorporated in SPARC, we are able to absorb both MHD and acoustic waves. A full description is provided in section 2.7.

2.7 Convolutional Perfectly Matched Layers (C-PMLs)[†]

Choosing boundary conditions that accurately satisfy user requirements in numerical simulations represents a constant challenge. In calculations that involve waves, outgoing or radiating or absorbing conditions that allow for the clean removal of waves while ensuring the fidelity of the solution within the relevant computational domain (Colonius 2004) are commonly applied. A range of techniques such as the damping sponge (Colonius 2004; Lui 2003), radiating or outgoing conditions based on characteristics (Thompson 1990), a locally supersonic boundary-directed flow to advect waves out (Lui 2003), perfectly matched layers (PMLs, see Berenger 1994), etc. As can be anticipated, all these methods possess weaknesses ranging from mediocre absorption efficiency to long-term instabilities. One issue with characteristics-based (outgoing) boundary conditions is that they work best when waves are normal to the boundary. If waves strike the boundary at large angles, a significant reflection typically results. In contrast, sponges may be superior at damping waves of all angles and are easy to implement in code. However, they are computationally expensive and the improvement in absorption may not be sufficient to justify the additional incurred cost.

The criterion of high-fidelity absorption is possibly satisfied best by PMLs, originally developed by Berenger (1994) to absorb outgoing electromagnetic waves. The primary concept is to perform a local analytic continuation of the (real) wave vector into the complex plane in the vicinity of the boundary. Wave vectors that are pointed in the outward direction and perpendicular to the boundary take on an imaginary coefficient, which can act to substantially damp outgoing waves in the PML. Discretization errors apart, the PML can be highly effective (Berenger 1994) and has subsequently set off a minor scientific industry in matched-layer construction. Discretization is known to introduce weak reflections but among absorbing boundary formulations, PMLs remain a computationally lean and highly effective choice. However, it was realized that in the split formulation introduced by Berenger (1994), absorption efficiency can decrease at large-angle or grazing incidence (Collino and Monk 1998; Winton and Rappaport 2000). Additionally waves at grazing incidence can destabilize the numerical method.

Since the original work of Berenger, there have been significant advances in constructing accurate and stable PMLs. Convolutional PMLs or CPMLs as they are termed (Roden, J. A. and Gedney, S. D. 2000; Festa and Vilotte 2005; Komatitsch and Martin 2007), have an additional Butterworth filter that acts to increase absorption efficiency at grazing incidence. This formalism has been applied to simulations of wave propagation in anisotropic elastic media by Komatitsch and Martin (2007) and the stability of the method has been characterized by Komatitsch and Martin (2007) and Meza-Fajardo and Papageorgiou (2008). Here we discuss the C-PML for the 3-D linearized ideal MHD equation in stratified media (Hanasoge et al 2010b).

[†] Content largely derived from Hanasoge et al (2010b)

The use of PMLs for MHD is not new; codes developed by Khomenko and Collados (2006) and Parchevsky and Kosovichev (2007), utilize the classical split-PML formulation to solve the ideal MHD wave equation in stratified media. There are drawbacks to these implementations however. For instance, Khomenko and Collados (2006) note the appearance of instabilities on long integration times. The technique that Parchevsky and Kosovichev (2007) apply involves the introduction of a small arbitrary constant that acts possibly as a sponge and diminishes the performance of their PML.

2.7.1 Constructing the Matched Layer

To begin, take the scenario of a wave propagating towards the upper boundary (the positive z direction in our convention). Leaving aside the stratification of the background medium, a simple traveling wave expansion gives us $v_z \sim A e^{i(k_z z - \omega t)}$, where A is wave amplitude, k_z is the wavenumber in the z direction, ω is temporal frequency, and t is time. Berenger (1994) suggests that the following transformation be applied to k_z:

$$k_z \rightarrow k_z \left[1 - \frac{d}{i\omega} \right],$$ (2.14)

where $d = d(z)f(x,y,z_0) \geq 0$ is some (damping) parameter and $z = z_0$ is the vertical location corresponding to the start of the CPML. The term $f(x,y,z_0)$ is introduced to capture large variations in wavespeed in the lateral directions, i.e. non-PML, at $z = z_0$. It is useful to note that the transformation is applied only to k_z, whereas wavenumbers k_x and k_y are retained as is. The matched-layer formulation in equation (2.14) is known to be unstable in a variety of scenarios (i.e., with $f(x,y,z_0) = 1$), specifically when mean flows are non-zero (i.e., $\mathbf{v}_0 \neq \mathbf{0}$, see Hu 2001; Appelö, D. et al 2006), and for anisotropic media (Bécache et al 2003; Komatitsch and Martin 2007; Khomenko 2009). A stable PML formulation for Maxwell equations was first discussed by Roden, J. A. and Gedney, S. D. (2000) and extended to solve the linear elastic wave equation by Festa and Vilotte (2005) and Komatitsch and Martin (2007). They used

$$k_z \rightarrow k_z \left[\kappa + \frac{d}{\alpha - i\omega} \right],$$ (2.15)

where $\alpha = \alpha(z) > 0$ and $\kappa = \kappa(z) \geq 1$ are new parameters. With the inclusion of α, a filtering term, the problem of absorbing waves arriving at grazing incidence at the boundary was significantly mitigated, also helping to stabilize the numerical method. Here, we extend the technique described by Komatitsch and Martin (2007) to linearized MHD. Basically, the transformation described in equation (2.15) is a spatio-temporal stretching of the grid in the z direction

$$\tilde{z} = \left(\kappa + \frac{d}{\alpha - i\omega} \right) z.$$ (2.16)

Thus vertical derivatives in the PML region must be calculated in terms of this (new) stretched coordinate. The derivative of a function $\psi(x,y,z,t)$ in the vertical z direction is therefore given by

$$\partial_z\psi \rightarrow \partial_{\tilde{z}}\psi, \tag{2.17}$$

where $\partial_{\tilde{z}}\psi$ is

$$\partial_{\tilde{z}}\psi = \left[\frac{1}{\kappa} - \frac{d/\kappa^2}{(d/\kappa+\alpha)-i\omega}\right]\partial_z\psi. \tag{2.18}$$

It is seen that the first term within the square brackets on the right-hand side of equation (2.18) is just the conventional derivative divided by κ. However, the second term requires more nuanced handling since it is a product in temporal Fourier domain, i.e. it is a convolution in time of multiplicative factor with the vertical derivative of ψ. To evaluate this temporal convolution, Roden, J. A. and Gedney, S. D. (2000); Festa and Vilotte (2005); Komatitsch and Martin (2007) apply a recursion formula at each timestep. Here, we introduce auxiliary variables described by attendant differential equations to evaluate $\chi(x,y,z,\omega) = \bar{s}\partial_z\psi$, where

$$\bar{s}(x,y,z,\omega) = -\frac{d/\kappa^2}{(d/\kappa+\alpha)-i\omega}. \tag{2.19}$$

We choose χ and ψ such that $\chi(x,y,z,t \leq 0) = 0$ and $\partial_z\psi = 0$ for all $t \leq 0$. Introducing the following ansatz for χ

$$\partial_t\chi = -\frac{d}{\kappa^2}\partial_z\psi - \left(\frac{d}{\kappa}+\alpha\right)\chi, \tag{2.20}$$

ensures that it has the required temporal frequency response,

$$\chi(x,y,z,\omega) = -\frac{d/\kappa^2}{(d/\kappa+\alpha)-i\omega}\partial_z\psi. \tag{2.21}$$

Thus the equations for the C-PML are

$$\rho = -\nabla_h\cdot(\rho_0\boldsymbol{\xi}) - \rho_0\partial_{\tilde{z}}\xi_z - \xi_z\partial_{\tilde{z}}\rho_0, \tag{2.22}$$

$$\rho_0\partial_t^2\boldsymbol{\xi} = \nabla_h\cdot\left[\frac{\mathbf{B}\mathbf{B}_0}{4\pi} + \frac{\mathbf{B}_0\mathbf{B}}{4\pi} - \left(p + \frac{\mathbf{B}_0\cdot\mathbf{B}}{4\pi}\right)\mathbf{I}\right]$$

$$+ \partial_{\tilde{z}}\left[\frac{B_z\mathbf{B}_0}{4\pi} + \frac{B_{0z}\mathbf{B}}{4\pi} - \left(p + \frac{\mathbf{B}_0\cdot\mathbf{B}}{4\pi}\right)\mathbf{e}_z\right]$$

$$- \rho\tilde{g}_0\mathbf{e}_z - \sigma\rho_0\boldsymbol{\xi} + \mathbf{S}, \tag{2.23}$$

$$p = -c_0^2\rho_0\nabla_h\cdot\boldsymbol{\xi} - c^2\rho_0\partial_{\tilde{z}}\xi_z - \boldsymbol{\xi}\cdot\nabla_h p_0 - \xi_z\partial_{\tilde{z}}p_0, \tag{2.24}$$

$$\mathbf{B} = \nabla_h\cdot(\mathbf{B}_0\boldsymbol{\xi} - \boldsymbol{\xi}\mathbf{B}_0) + \partial_{\tilde{z}}(B_{0z}\boldsymbol{\xi} - \xi_z\mathbf{B}_0), \tag{2.25}$$

where $\nabla_h = \mathbf{e}_x\partial_x + \mathbf{e}_y\partial_y$ are lateral derivatives (non-C-PML directions), and $\tilde{g}_0, \partial_{\tilde{z}}p_0$, $\partial_{\tilde{z}}\rho_0$ are the modified background gravity, pressure, and density gradients in the absorption layer, respectively, and $\mathbf{I}$ is the unit dyad that satisfies $\{\mathbf{I}\}_{ij} = \delta_{ij}$. The term $-\sigma\rho_0\boldsymbol{\xi}$, where $\sigma = \sigma(x,y,z)$ is inserted in the momentum equation because it is seen to stabilize the numerical evolution. We note that we are uninterested in the solution within the absorbing layer and discard it when evaluating the accuracy of numerical results. Therefore the modified equations, with altered stratification and not enforcing $\nabla \cdot \mathbf{B} = 0$, etc. are justified on the basis that the solution within the relevant computational domain is accurate and the numerical method, stable. Equations (2.22) through (2.25) in addition to Equation (2.20) give

$$\rho = -\nabla_h \cdot (\rho_0 \boldsymbol{\xi}) - \rho_0 \left[\frac{1}{\kappa}\partial_z\xi_z + \Psi\right] - \xi_z\partial_{\bar{z}}\rho_0, \tag{2.26}$$

$$\partial_t\Psi = -\frac{d}{\kappa^2}\partial_z\xi_z - \left(\frac{d}{\kappa}+\alpha\right)\Psi, \tag{2.27}$$

$$\partial_{\bar{z}}\rho_0 = \frac{\alpha/\kappa}{d/\kappa+\alpha}\partial_z\rho_0, \tag{2.28}$$

$$\rho_0\partial_t^2\boldsymbol{\xi} = \nabla_h \cdot \left[\frac{\mathbf{B}\mathbf{B}_0}{4\pi} + \frac{\mathbf{B}_0\mathbf{B}}{4\pi} - \left(p+\frac{\mathbf{B}_0\cdot\mathbf{B}}{4\pi}\right)\mathbf{I}\right]$$
$$+ \frac{1}{\kappa}\partial_z\left[\frac{B_z\mathbf{B}_0}{4\pi} + \frac{B_{0z}\mathbf{B}}{4\pi} - \left(p+\frac{\mathbf{B}_0\cdot\mathbf{B}}{4\pi}\right)\mathbf{e}_z\right]$$
$$+ \boldsymbol{\theta} - \rho\tilde{g}_0\mathbf{e}_z - \sigma\rho_0\boldsymbol{\xi} + \mathbf{S}, \tag{2.29}$$

$$\partial_t\boldsymbol{\theta} = -\frac{d}{\kappa^2}\partial_z\left[\frac{B_z\mathbf{B}_0}{4\pi} + \frac{B_{0z}\mathbf{B}}{4\pi} - \left(p+\frac{\mathbf{B}_0\cdot\mathbf{B}}{4\pi}\right)\mathbf{e}_z\right]$$
$$- \left(\frac{d}{\kappa}+\alpha\right)\boldsymbol{\theta}, \tag{2.30}$$

$$\tilde{g}_0 = \frac{\alpha/\kappa}{d/\kappa+\alpha}g_0, \tag{2.31}$$

$$p = -c_0^2\rho_0\nabla_h\cdot\boldsymbol{\xi} - c^2\rho_0\left[\frac{1}{\kappa}\partial_z\xi_z + \Psi\right]$$
$$- \mathbf{v}\cdot\nabla_h p_0 - v_z\partial_{\bar{z}}p_0, \tag{2.32}$$

$$\partial_{\bar{z}}p_0 = \frac{\alpha/\kappa}{d/\kappa+\alpha}\partial_z p_0, \tag{2.33}$$

$$\mathbf{B} = \nabla_h \cdot (\mathbf{B}_0\boldsymbol{\xi} - \boldsymbol{\xi}\mathbf{B}_0)$$
$$+ \frac{1}{\kappa}\partial_z(B_{0z}\boldsymbol{\xi} - \xi_z\mathbf{B}_0) + \boldsymbol{\eta}, \tag{2.34}$$

$$\partial_t\boldsymbol{\eta} = -\frac{d}{\kappa^2}\partial_z(B_{0z}\boldsymbol{\xi} - \xi_z\mathbf{B}_0) - \left(\frac{d}{\kappa}+\alpha\right)\boldsymbol{\eta}. \tag{2.35}$$

We recall that the vector memory (auxiliary) variables $\boldsymbol{\theta},\boldsymbol{\eta}$ are necessary for calculating the temporal convolution in equation (2.18) and that $\boldsymbol{\eta} = (\eta_x,\eta_y,0)$.

We see that the stratification is altered within the absorbing region. To understand this, consider the derivative $\partial_{\bar{z}}p_0$. The differential equation (2.20) when applied to the temporally constant $\partial_z p_0$ leads to

$$\partial_{\bar{z}} p_0 = \frac{1}{\kappa} \partial_z p_0 + \chi, \tag{2.36}$$

$$\partial_t \chi = -\frac{d}{\kappa^2} \partial_z p_0 - \left(\frac{d}{\kappa} + \alpha\right) \chi. \tag{2.37}$$

Because the background is time invariant, we set the time derivative in Equation (2.37) to zero,

$$\partial_{\bar{z}} p_0 = \frac{\alpha/\kappa}{d/\kappa + \alpha} \partial_z p_0. \tag{2.38}$$

Similarly, the gradient of density in the CPML region $\partial_{\bar{z}} \rho_0$ can also be evaluated. In order to maintain hydrostatic support within the CPML, we modify gravity by multiplying it with the (same) factor, $\alpha/(d + \kappa\alpha)$, thereby arriving at Equations (2.28), (2.31), and (2.33).

In the presence of magnetic fields, three wave branches with distinct dispersion relations are known to exist slow, fast, and Alfvén (these branches become degenerate when the Alfvén and sound speeds are the same). Wave propagation in magnetic media is anisotropic, especially when the Alfvén is much larger than the sound speed (Goedbloed and Poedts 2004). The primary difference between these branches is that Alfvén waves are incompressible transversely propagating shear-like oscillations whereas slow and fast modes are *magneto-acoustic* in character, i.e. they are magnetically guided and compressive. The MHD dispersion relations exhibit pathologies in that at specific wavenumbers, the phase and group speeds become arbitrarily small (pp. 195–214 of Goedbloed and Poedts 2004). This renders the calculation unstable in and around the entry of the absorption region. To stabilize the calculation, we introduce the sponge-like $-\sigma\rho_0 \xi$ term to the momentum equation in regions where the magnetic field is non-zero. We show an instance of a wave guided by a magnetic flux tube in Figure 2.3. Horizontal motions shake the flux tube at the lower boundary thereby exciting MHD waves; the fast modes (as the name suggests) arrive at the upper boundary first, followed by a dispersing train of slow and Alfvén modes, some parts of which can take hours (or longer) to reach the upper boundary. The Alfvén wavespeed is directly proportional to magnetic field strength, thus they propagate rapidly in strongly magnetized regions and conversely, slowly in weak field.

The choices for free parameters κ, d, α can have a significant impact on the efficiency of the absorption. Along the lines of Komatitsch and Martin (2007), we set $\alpha = \pi f_0$, where f_0 is the characteristic frequency of the waves. The term α falls linearly to zero over the length of the absorption layer, from its maximum at the start of the layer to zero at the boundary. Conversely, the damping function, $d = f(x, y, z_0) (z/L)^N$, is zero at the layer entry and reaches its maximum value at the boundary. Through trial and error, we determine that for the Euler equations, choosing the following for f induces efficient absorption

$$f(x, y, z_0) = -\frac{N+1}{2L} \bar{c} \log_{10} R_c, \tag{2.39}$$

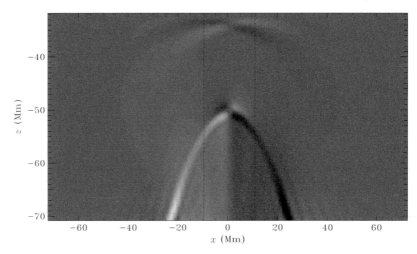

Fig. 2.3 Snapshot as seen in vertical velocity of MHD waves propagating along a magnetic flux tube. The horizontal boundaries of the flux tube, i.e. the full-width at half maximum of the field strength, are marked by dot-dash lines. Waves at the vanguard, i.e., those closest to the upper boundary are fast modes propagating at $\sqrt{c_0^2 + c_A^2}$, where $c_A = ||\mathbf{B}||/\sqrt{4\pi\rho_0}$ is the Alfvén speed. A dispersed train of slow and Alfvén modes follows, propagating at a variety of slower speeds. The fastest speeds occur at the centre of the tube where the field is strongest, whereas speeds at the edge where the field is weak ($x = \pm 30$ Mm) propagate very slowly (Hanasoge et al 2010b; reproduced with permission ©ESO).

where L is layer thickness, R_c is a tolerance limit on the amount of reflection, and $\bar{c}$ (not a function of z) is the characteristic sound speed in the layer. Here we use

$$f(x,y,z_0) = -\frac{N+1}{2L}c_w \log_{10} R_c, \qquad (2.40)$$

where $c_w(x,y,z_0) = \sqrt{c_0(x,y,z_0)^2 + c_A(x,y,z_0)^2}$ is the fastest propagation speed in the layer (set at vertical location $z = z_0$). The wave propagation speed is substantially different in strong field regions from corresponding non-magnetic areas. We therefore allow c_w to be a function of horizontal coordinates, i.e., $c_w = c_w(x,y,z_0)$. We denote sound and Alfvén speeds by c and $c_A = ||\mathbf{B}||/\sqrt{4\pi\rho_0}$, respectively. Finite values of κ are used to absorb (non-propagating) evanescent waves (Roden, J. A. and Gedney, S. D. 2000; Bérenger 2002). Indeed, while our simulations produce evanescent waves, we find by trial that when κ varies from 1 to 8 over the layer, the calculation goes unstable. For smaller values of κ, absorption efficiency is not appreciably different from when $\kappa = 0$. We therefore choose $\kappa = 1$ over the entire layer. For the sponge, we choose $\sigma = -[(N+1)c_A/L](z/L)^N \log_{10} R_c$. With the introduction of this sponge, it is a valid question as to whether this formulation is perfectly matched. However, because $\sigma \propto c_A$ the technique continues to be perfectly matched for the stratified Euler equations, i.e. when $c_A = 0$ (zero-magnetic field).

Equations (2.26) to (2.35) show that a total of 6 auxiliary variables have been introduced: for v_z, the three (vector) Lorentz-force components, and two in the induction equation. The overhead on memory and computation is small: (1) typically, 8–10 grid points are sufficient to accommodate the CPML; we have to store the six memory variables only over a limited number of grid points, and (2) additional computation is limited to a small number of multiplications and additions, as required for evolving the auxiliary differential equations (2.27), (2.30), and (2.35).

2.7.2 Numerical results

2.7.3 Waves in a non-magnetic stratified fluid

We first investigate the absorption efficiency of the C-PMLs. We design a stratified polytrope with index $m = 2.15$ and set it to be the background medium. The vertical extent of the computational domain is thick enough to accommodate some 2.6 density scale heights and 3.72 pressure scale heights (e.g., the pressure at the bottom of the domain is $e^{3.72} = 41.5$ times the value at the top). The full properties of the polytrope are described in Hanasoge et al (2010b). Waves are excited in a spatially localized region 18 Mm above the bottom boundary (the vertical extent of the computational domain is 68 Mm). For this calculation, we choose $N = 2, R_c = 0.1\%, f_0 = 0.005$ Hz and the upper and lower C-PMLs are each 10 grid points thick. We display snapshots from the calculation at four different instants of time in Figure 2.4. Because the scale on all plots is held constant, and it can be seen that the upper CPML almost completely absorbs the incident waves. To quantify the extent and fidelity of the absorption, we track the temporal evolution of the wave-energy invariant (summed over the entire grid) in Figure 2.5, (Bogdan et al 1996)

$$e = \frac{1}{2}\rho_0 ||v||^2 + \frac{p^2}{2\gamma p_0}, \qquad (2.41)$$

where $v = \partial_t \xi$.

Next we study the stability of the C-PMLs by integrating the wave equation over some 300 wave periods (maximum wave frequency ~ 6 mHz; 12-hour integration). To perform this test, we use a more solar-like (realistic) setup. The computational domain now includes the solar photosphere, where both density and pressure reduce exponentially rapidly with height. In total, the domain spans 21 density scale heights, i.e. the ratio between the bottom and top densities is around 1.3 billion. The stratification is discussed in, e.g., Hanasoge et al (2008).

The computational box is of size $200 \times 200 \times 35$ Mm^3 (two lateral dimensions times vertical length) and the grid has $256 \times 256 \times 300$ points. Vertically, the box spans 35 Mm, extending from 34 Mm below the solar photosphere to 1 Mm into the atmosphere. Waves are stochastically excited in the Sun by vigorous near-surface turbulence (e.g., Stein and Nordlund 2000). To replicate this, we add a laterally

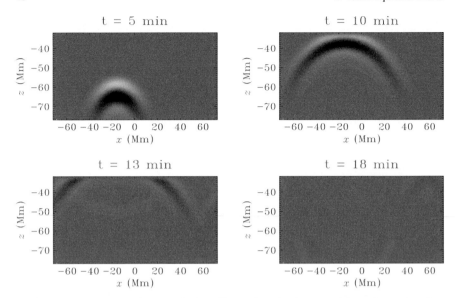

Fig. 2.4 Normalized vertical wave velocity $\sqrt{\rho_0}v_z$ at four time instants. All panels are plotted on an identical grey scale, and it is seen that waves in the $t = 18$-minutes picture are almost invisible (Hanasoge et al 2010b; reproduced with permission ©ESO).

statistically uniform random forcing function, localized to a shallow depth, to the vertical momentum equation $\mathbf{S} = S(x,y,z,t)\mathbf{e}_z$.

Over the time period of integration, we find the calculation to be stable. Figure 2.6 shows the spatio-temporal power spectrum of the vertical velocity component v_z extracted at a height $z = 200$ km above the photosphere. The spectrum is obtained by spatially and temporally Fourier transforming the raw velocity time series and summing over the absolute value squared as a function of total wavenumber. The horizontal axis is the non-dimensional lateral wavenumber $k_h R_\odot$, where k_h is the lateral wavenumber, and the vertical axis is the frequency in mHz.

We also compare the performance of CPMLs (Figure 2.6) and damping sponges (Figure 2.7). Typically, weak reflections from the lower boundary produce artifacts in the modal power spectrum, with the normal modes responding to the finiteness of the computational box. This is manifested in the flattening of the modal ridges, seen in Figure 2.7 but not in Figure 2.6 where wave absorption at the lower boundary is more efficient.

2.7.4 Stratified MHD fluid

Our first test is similar to that shown in Figure 2.5. The simplest magnetic configuration is a constant uniform field; we thus embed a constant inclined magnetic field in the polytrope (see Hanasoge et al 2010b). We choose the field strength such

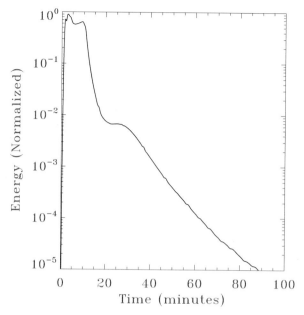

Fig. 2.5 Time evolution of the normalized modal energy as computed with equation (2.41). The initial (small) drop in energy is due to the arrival of waves at the lower C-PML from the source. The large drop in the energy corresponds to the first arrivals at the upper C-PML. Modes with large horizontal wavenumbers and weakly reflected waves arrive gradually at the boundaries at later times, resulting in an extended energy decay (Hanasoge et al 2010b; reproduced with permission ©ESO).

that the maximum Alfvén speed is approximately four times the sound speed (at the upper boundary). We horizontally shake the field lines at the upper boundary to excite MHD waves. The fast and Alfvén waves are the first to arrive at the upper boundary, followed by a long dispersive train of slow modes. Because of this prolonged arrival, it is more difficult to demonstrate the absorptive properties of the CPML than in Figure 2.5. Moreover, the choice for energy invariants for MHD waves in stratified media is not easily made. Various authors who have studied this issue (Bray and Loughhead 1974; Parker 1979; Leroy 1985) arrive at differing versions (personal communication, P. S. Cally 2009). Here, we use the following form (kinetic, thermal, and magnetic energies, respectively, Bray and Loughhead 1974):

$$e = \frac{1}{2}\rho_0 ||\boldsymbol{v}||^2 + \frac{p^2}{2\gamma p_0} + \frac{||\mathbf{B}||^2}{8\pi}. \tag{2.42}$$

The time evolution of the total energy (calculated over the entire domain) is plotted in Figure 2.8.

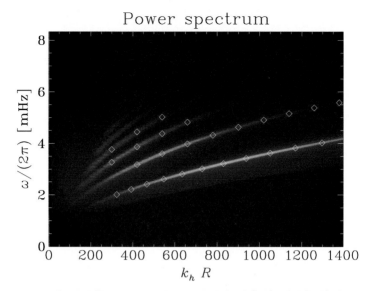

Fig. 2.6 Contour plot of vertical velocity power spectrum measured at of $z = 200$ km above the photosphere from a 12-hour long simulation. The upper and lower boundaries are lined with CPMLs. The horizontal axis is normalized wavenumber, the vertical is temporal frequency, and contours of high power represent resonant modes of the computational box. Symbols on top of the power contours are values of the resonant frequencies calculated using MATLAB's boundary-value problem solver `bvp4c` (Hanasoge et al 2010b; reproduced with permission ©ESO).

In this last experiment, we attempt to highlight the long-term stability of the method. In the calculation, we excite waves in non-magnetic regions which then propagate through a magnetic flux tube that is placed in the stratified polytrope (for details on stratification, see Hanasoge et al 2010b; Moradi et al 2009). The magnetic field configuration is pictorially displayed in Figure 2.9. Over a temporal integration period of some 150–300 wave periods (i.e., 12 hours), the system showed numerical stability. An accurate numerical scheme must maintain $\nabla \cdot \mathbf{B} = 0$ (Tóth 2000). Among other reasons, discretization errors can be a source of finite $\nabla \cdot \mathbf{B}$. In addition, the boundary formulation does not explicitly conserve $\nabla \cdot \mathbf{B} = 0$. It is therefore necessary to quantify and determine the growth of error in maintaining divergence-free magnetic field fluctuations. Here we introduce a normalized measure to estimate the departure from zero divergence:

$$n_e = \frac{L_2 \left[\int_{z_1}^{z_2} dz ||\nabla \cdot \mathbf{B}|| \right]}{L_2 \left[||\mathbf{B}|| \right]}, \tag{2.43}$$

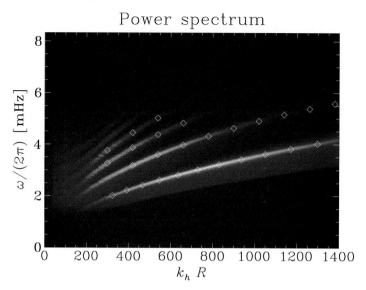

Fig. 2.7 Vertical velocity power spectrum measured at of $z = 200$ km above the photosphere from a 12-hour long simulation. The upper and lower boundaries are lined with damping sponges, which are lower in efficiency than CPMLs. Waves reflect weakly off the lower boundary causing the dispersion relation to change in curvature at sufficiently large values of $v/(k_h R_\odot)$. The axes are as in Figure 2.6. The symbols mark resonant mode frequencies as computed using MATLAB (Hanasoge et al 2010b; reproduced with permission ©ESO).

where z_1, z_2 are the boundaries of the relevant portion of the computational domain. The L_2 norm of a function $f(x)$ sampled on a discrete grid $\{x_i\}$ is defined as

$$L_2[f(x)] = \sqrt{\sum_i f(x_i)^2}. \tag{2.44}$$

The time history of n_e is displayed in Figure 2.10 (left). Over the period of the calculation, n_e is essentially invariant, possibly decreasing with time. The evolution of the numerator and denominator of n_e is shown in the right panel of Figure 2.10. Wave excitation via the source (**S**) term in the vertical momentum equation excites magnetic waves, resulting in a secular increase in the magnetic energy.

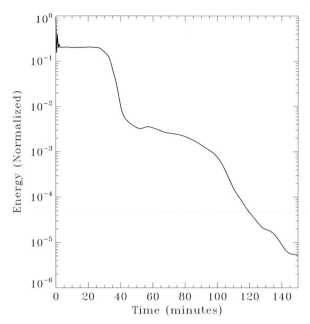

Fig. 2.8 Temporal history of MHD wave energy as defined in equation (2.42), summed over the entire grid. Wave dispersion and substantial differences in the wavespeeds between fast, slow, and Alfvén modes result in an extended energy decay (compare with Figure 2.5). The first drop at $t \sim 20 - 40$ min is timed with fast-mode arrivals at the upper and lower absorption layers. The subsequent gradual decay is related to downward-propagating slow and Alfvén modes and their eventual arrival at the lower absorption layer (Hanasoge et al 2010b; reproduced with permission ©ESO).

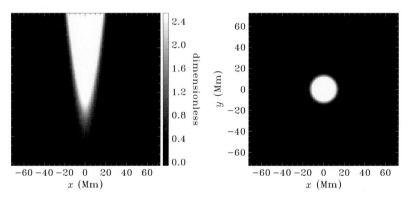

Fig. 2.9 Pictorial description of the magnetic field used in the long-time MHD integration test. The left panel shows the ratio of the local Alfvén to the sound speed as a function of lateral coordinate x and vertical coordinate z. The field is dynamically irrelevant at depth because hydrodynamic pressure vastly exceeds magnetic pressure. A horizontal cut through the fast mode speed distribution ($\sqrt{c_A^2 + c_0^2}$) taken at the solar photosphere is shown on the right panel (Hanasoge et al 2010b; reproduced with permission ©ESO).

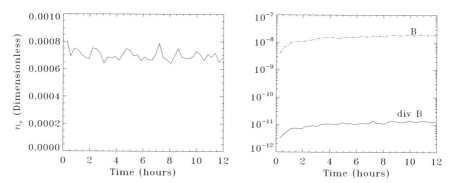

Fig. 2.10 The error measured as deviations from $||\boldsymbol{\nabla} \cdot \mathbf{B}|| = 0$ using n_e (Eq. [2.43]) is shown on the left panel. The time histories of the L_2 norms of $\int_{z_1}^{z_2} dz \, ||\boldsymbol{\nabla} \cdot \mathbf{B}||$ (solid) and $||\mathbf{B}||$ (dot-dash) are shown on the right panel (Hanasoge et al 2010b; reproduced with permission ©ESO).

Chapter 3
Adjoint Optimization[‡]

Stochastic wavefields, present in a variety of media, such as stars and planets, may be created by the action of random sources of wave excitation, where source location, amplitude, and phase are random variables. Without knowledge of the exact realization of all relevant sources of wave excitation, raw time series of wavefield velocities contain no useful seismic information. However, it was discovered that seismically relevant data were contained in time-averages over many source realizations of second-order correlations of wavefield velocities in the Sun (Duvall et al (1993); for noise tomography, see, Shapiro and Campillo 2004). These correlations contain components of noise, whose standard deviation diminishes as the inverse square root of the temporal length of averaging, which is the consequence of wave excitation by a stationary random process (see, e.g., Gizon and Birch 2002; Gizon 2004). In the Sun, turbulent convection, driven by radiative thermal losses at the surface, is the cause of wave generation. This random process is adequately represented by a stationary and laterally homogeneous random process (Gizon 2004). Typically, the correlation time of solar convection (granulation) is 10 min and the correlation length is 1 Mm. While the correlation time is on the order of the wave period, the correlation length is smaller than the wavelength and thus the assumption of spatially uncorrelated sources is reasonable. The response of the Sun to excitation by turbulent convection produces a power spectrum that peaks near 3 mHz.

Woodard (1997) and Gizon and Birch (2002) were among the first to utilize these ideas towards the construction of a theoretical description of helioseismic measurements. A prescription to compute sensitivity kernels and model excitation noise was described by Gizon and Birch (2002); Gizon (2004). Various authors, e.g., Birch et al (2004), Birch and Gizon (2007), and Jackiewicz et al (2007), subsequently used this theory to derive sensitivity kernels for flows and sound-speed perturbations for translationally invariant (laterally homogeneous) background models. The basic recipe described in Gizon and Birch (2002) to compute travel-time sensitivity

---

[‡] Material in this section is taken primarily from Hanasoge et al (2011) and Hanasoge et al (2012a).

© The Author 2015
S. Hanasoge, *Imaging Convection and Magnetism in the Sun*, SpringerBriefs in Mathematics, DOI 10.1007/978-3-319-27330-3_3

kernels for randomly excited waves was general; however no attempt was made to relate this method to the adjoint method, which enables computation of kernels for heterogeneous background models using numerical wave simulations.

The adjoint method has a long history (Lions 1971) and is widely used in fluid control (e.g., Bewley et al 2001; Giles and Pierce 2000; and references therein), airfoil optimization (e.g., Jameson 1988), meteorology (e.g., LeDimet and Talagrand 1986; Talagrand and Courtier 1987), global helioseismology (Rosenwald and Rabaey 1991), and terrestrial seismology (e.g., Tarantola 1984; Tromp et al 2005, 2010). Real-world optimization problems are typically functions of large numbers of parameters, ill posed, and computationally expensive. For instance, one may envisage the difficulty in minimizing drag due to flow over an airfoil or seeking a model of Earth's interior that optimally fits observed seismograms, simply due to large number of ways one may alter the system. What parameters should one vary in order to achieve optimality? It is evident that the gradient of the misfit function with respect to various parameters tells us how to march towards a stationary point, i.e., a point at which the derivative of a quantity vanishes. The adjoint method provides an algorithm to compute Fréchet derivatives and hence the gradient with relatively small computational expense.

In this chapter, we discuss the extension of the adjoint method to the computation of helioseismic sensitivity kernels relative to arbitrarily heterogeneous background models. The complexity of equations in the presence of strong lateral inhomogeneity is such that evaluation of kernels must proceed by computational means. A remarkable outcome of allowing for lateral (horizontal) variations in the background model is the ability to compute vector kernels for magnetic fields. Since the perturbation induced by fields scales as $O(|\mathbf{B}|^2)$, where $\mathbf{B}$ is the background field, and the action of the Lorentz force is anisotropic, it is non-trivial to derive magnetic field kernels about a 1-D solar model (i.e., where properties are only a function of one coordinate, the radius). However, if we were to linearize around a 3-D background model that contains an embedded field $\mathbf{B}$, kernels describing shifts in helioseismic measurements due to small vector variations in the magnetic field emerge naturally.

Evaluating kernels in the context of helioseismology requires the computation of six wavefields per measurement. Losing the luxury of being able to translate kernels from one horizontal position to another therefore comes at a stiff computational price, since one must, in principle, evaluate kernels around each observational pixel, an impossible feat in helioseismology owing to the vast numbers of observations. MDI records velocities at approximately 1 million points on the solar photosphere, and with the advent of SDO, HMI now captures velocities at more than 16 million pixels every 45 seconds. Computing kernels at all these points is neither computationally feasible nor is it clear that there is sufficient independent information to require such a massive calculation. Consequently, we introduce the concept of "master pixels," a finite constellation of points which we consider interesting enough to invest this sizeable computational effort. However, once a number of these pixels have been chosen, *every* cross correlation measurement, one of whose antennae is a master pixel, may be utilized in the inversion without affecting computational cost (Tromp et al 2010).

Sunspots are substantial deviations from the quiet Sun, with umbral temperatures dropping by as much as 20% from ambient conditions. Numerous questions swirl around sunspot physics, such as understanding their long-time stability (compared to convective turnover timescales) and appreciating their creation, emergence, and eventual death. The use of helioseismic waves to probe the structure of sunspots has a long and controversial history (for a review, see, e.g., Gizon et al 2010). Inversions for sunspot sub-surface structure and dynamics (e.g., Kosovichev and Duvall 1997) attempt to explain away the observed effects on seismic waves by an entirely isotropic wavespeed, an approximation that has faced subsequent marginalization (e.g., Gizon et al 2009) owing to the widespread recognition of strong anisotropies prevalent in sunspots. Forward modeling of wave propagation in sunspots has generated a deeper appreciation for measurements and fully realistic non-linear sunspot evolution calculations (Rempel et al 2009) have proven successful. However, posing an inverse problem that accounts for these anisotropies remains an outstanding problem of great relevance, towards whose eventual resolution the computation of kernels is a significant step.

In this chapter, we shall primarily discuss the mathematical underpinnings of the adjoint method and its applicability to helioseismology. A computational algorithm to implement the analysis is described. Perturbations, such as sunspots, are significant deviations from the quiet Sun and shifts in helioseismic measurements in and around sunspots are substantial and unlikely to scale linearly with perturbation strength (when measured relative to the quiet Sun). A means of carrying out iterative inversions in such situations is described. With the increasing availability of computational resources, demand for greater accuracy in the interpretation of helioseismic measurements, and the advent of higher-quality observations, the introduction of such a technique is thought to be timely.

3.1 Governing equations of the helioseismic wavefield

We reproduce the magneto-hydrostatic equilibrium relation equation (2.1) that governs the background state

$$\nabla p = \rho \mathbf{g} + (\nabla \times \mathbf{B}) \times \mathbf{B}, \tag{3.1}$$

$$\nabla \cdot \mathbf{g} = -4\pi G \rho, \tag{3.2}$$

where p is the background pressure, ρ the density, $\mathbf{B}$ the magnetic field, $\mathbf{g} = -g\mathbf{e}_r$ gravity, G the universal gravitational constant, and $\mathbf{e}_r$ the radially outward unit vector (e.g., Lynden-Bell and Ostriker 1967; Goedbloed and Poedts 2004; Cameron et al 2007). In this formalism, we consider background flows ($\mathbf{v}$) to be too weak to contribute significantly towards maintaining equilibrium, and hence we neglect advection-related forces in equation (3.1). This implies that $||\mathbf{v}|| \ll \sqrt{gL}$, where L is the characteristic flow length scale, and that Lorentz forces are primarily balanced by pressure gradients and gravity. However any flows that are present must satisfy the continuity equation, and we require therefore that $\nabla \cdot (\rho \mathbf{v}) = 0$. The magnetic permeability constant $4\pi\mu_0$ has been absorbed into the definition of the field. We do not keep rotation terms in the force balance equation because Coriolis and centrifugal forces are five orders in magnitude smaller than surface gravity. We also invoke the Cowling approximation, allowing us to ignore changes in the gravitational potential induced by wave motions. Small-amplitude wave propagation in a magnetic environment is described by the following dynamical wave operator (in temporal Fourier space; see Appendix A.3 for the convention)

$$\mathscr{L}\boldsymbol{\xi} = -\omega^2\rho\boldsymbol{\xi} - 2i\omega\rho\mathbf{v}\cdot\nabla\boldsymbol{\xi} - i\omega\rho\Gamma\boldsymbol{\xi} - \nabla(c^2\rho\nabla\cdot\boldsymbol{\xi}) - \nabla(\boldsymbol{\xi}\cdot\nabla p) + \mathbf{g}\nabla\cdot(\rho\boldsymbol{\xi})$$
$$- (\nabla\times\mathbf{B})\times[\nabla\times(\boldsymbol{\xi}\times\mathbf{B})] - \{\nabla\times[\nabla\times(\boldsymbol{\xi}\times\mathbf{B})]\}\times\mathbf{B}, \qquad (3.3)$$

where $\boldsymbol{\xi}$ is the displacement vector and c is the background sound speed. In order, terms on the right side denote acceleration (first term), flow advection, wave damping, pressure restoring forces (the term with $c^2\rho$), buoyancy (the next two terms), and magnetic Lorentz force (the final two), respectively. We assume that the upper boundary is placed far away from the solar photosphere and the wavefield satisfies zero-Dirichlet conditions (all fluctuations are zero on this bounding surface). The entire solar interior is enclosed within this volume. Following Gizon and Birch (2002) and Birch et al (2004), we mimic the complex frequency dependence of wave damping in the Sun by including the term $-i\omega\Gamma\boldsymbol{\xi}$, where Γ is the damping rate. We neglect second-order flow terms such as $\mathbf{v}\cdot\nabla(\mathbf{v}\cdot\nabla\boldsymbol{\xi})$; this is a reasonable approximation when the velocities roughly satisfy $||\mathbf{v}|| \ll \{\omega L, c, \sqrt{gL}\}$, where ω is the characteristic wave frequency. It may be verified that this inequality is satisfied for most solar phenomena (e.g., Birch and Gizon 2007). The full wave equation is given by $\mathscr{L}\boldsymbol{\xi} = \mathbf{S}$, where $\mathbf{S}(\mathbf{x},\omega)$ is a source term.

Flows and damping do not follow directly from the equilibrium equations. The emergence of wave damping is not well understood, and is thought to be due to a combination of the action of turbulence and radiation (e.g., Duvall et al 1998; Korzennik et al 2004); we are unable to realistically account for these phenomena and are therefore forced to introduce phenomenological damping terms. Solar flows, as discussed previously, are typically weak perturbations to the background. Further, constructing a background model with flows and magnetic fields is a remarkably difficult task (e.g., Beliën et al 2002). Such practical considerations have led us to introduce these terms in an ad-hoc fashion.

3.2 Minimizing misfit

A common optimization problem in helioseismology is that of reducing differences between observed and predicted travel times. Cross correlation amplitudes, which depend quasi-linearly on properties of the background model, are commonly measured but not typically used in inversions; conceptually, one may include these in the misfit with no additional effort (e.g., Gee and Jordan 1992; Fichtner et al 2008; Bozdağ et al 2011).

A convenient choice for the misfit function is the L_2 norm of these differences summed over a number of observation points

$$\mathscr{I}' = \frac{1}{2} \sum_{q,q'} \mathscr{N}_{qq'} [\tau_q^{(n)} - \tau_q^{\text{o}}][\tau_{q'}^{(n)} - \tau_{q'}^{\text{o}}], \tag{3.4}$$

where τ_q^{o} is the observed travel time, $\tau_q^{(n)}$ the predicted analog with (current) background model n, specified at points q, q', and $\mathscr{N}_{qq'}$ the inverse of the noise covariance between the two sets of measurements, assumed to be chi-squared distributed (Gizon 2004). Here the noise-covariance model is assumed to be stationary under changes of the background model, i.e., $\mathscr{N}_{qq'}$ does not change with iteration. Partial differential equation constrained optimization is the technique of minimizing this misfit with respect to a governing wave equation,

$$\mathscr{I} = \frac{1}{2} \sum_{q,q'} \mathscr{N}_{qq'} [\tau_q^{(n)} - \tau_q^{\text{o}}][\tau_{q'}^{(n)} - \tau_{q'}^{\text{o}}] - \int_\odot d\mathbf{x} \int d\omega \, \boldsymbol{\lambda} \cdot (\mathscr{L}\boldsymbol{\xi} - \mathbf{S}), \tag{3.5}$$

where $\boldsymbol{\lambda}$ is a Lagrange multiplier and the integration proceeds over all space $\mathbf{x}$ and frequency ω. As, e.g., Woodard (1997) and Gizon and Birch (2002) realized, a first step towards formal interpretation of measurements is to create functionals linking cross correlations and travel times to the input displacement field, i.e., to establish a relation of the form $\tau_q^{(n)} = \tau_q^{(n)}(\boldsymbol{\xi})$. For now, we choose to represent this in an abstract fashion, and in subsequent sections move towards greater detail. Let us posit that a change in misfit (3.4) may be written as

$$\delta\mathscr{I}' = \int_\odot d\mathbf{x} \int d\omega \, \mathbf{f}^\dagger \cdot \delta\boldsymbol{\xi}, \tag{3.6}$$

where $\mathbf{f}^\dagger$ is a function that connects variations in displacement field $\delta\boldsymbol{\xi}$, to those of travel-time misfit $\delta\mathscr{I}'$. Now changes in the misfit associated with the constrained problem (3.5) may be written as

$$\delta\mathscr{I} = \int_\odot d\mathbf{x} \int d\omega \, \mathbf{f}^\dagger \cdot \delta\boldsymbol{\xi} - \int_\odot d\mathbf{x} \int d\omega \, [\delta\boldsymbol{\lambda} \cdot (\mathscr{L}\boldsymbol{\xi} - \mathbf{S}) + \boldsymbol{\lambda} \cdot \delta\mathscr{L}\boldsymbol{\xi} + \boldsymbol{\lambda} \cdot \mathscr{L}\delta\boldsymbol{\xi}], \tag{3.7}$$

upon invoking (3.6) and setting $\delta\mathbf{S} = \mathbf{0}$. If the forward displacement field were to satisfy $\mathscr{L}\boldsymbol{\xi} = \mathbf{S}$, and we were able to eliminate terms involving $\delta\boldsymbol{\xi}$, then changes in the misfit would be functions only of $\boldsymbol{\lambda}$, $\boldsymbol{\xi}$, and the perturbed wave operator, which depends only on background properties such as sound speed, magnetic fields, density, etc. Now in order to accomplish this, we need to first be able to free $\delta\boldsymbol{\xi}$ from the action of the operator in the third term of equation (3.7). The property of adjointness or duality is central to such a manipulation. An operator $\mathscr{O}$ is said to be *self-adjoint* if it satisfies

$$\int_\odot d\mathbf{x} \, \boldsymbol{\lambda} \cdot \mathscr{O}\boldsymbol{\xi} = \int_\odot d\mathbf{x} \, \boldsymbol{\xi} \cdot \mathscr{O}\boldsymbol{\lambda}. \tag{3.8}$$

For the boundary conditions chosen here, it may be demonstrated that the ideal MHD operator, which contains no flow or dissipation terms, is an example (e.g., Goedbloed and Poedts 2004). However, the non-ideal operator (3.3) is not self-adjoint and obeys

$$\int_{\odot} d\mathbf{x}\, \boldsymbol{\lambda} \cdot \mathscr{L}\boldsymbol{\xi} = \int_{\odot} d\mathbf{x}\, \boldsymbol{\xi} \cdot \mathscr{L}^{\dagger}\boldsymbol{\lambda}, \tag{3.9}$$

where $\mathscr{L}^{\dagger}$, defined as adjoint to (3.3), is given by (see appendix A.1)

$$\mathscr{L}^{\dagger}\boldsymbol{\xi} = -\omega^2 \rho \boldsymbol{\xi} - i\omega\rho\Gamma\boldsymbol{\xi} + 2i\omega\rho\mathbf{v}\cdot\nabla\boldsymbol{\xi} - \nabla(c^2\rho\nabla\cdot\boldsymbol{\xi} + \boldsymbol{\xi}\cdot\nabla p) + \mathbf{g}\nabla\cdot(\rho\boldsymbol{\xi})$$
$$- [(\nabla \times \mathbf{B}) \times \{\nabla \times (\boldsymbol{\xi} \times \mathbf{B})\} + \{\nabla \times [\nabla \times (\boldsymbol{\xi} \times \mathbf{B})]\} \times \mathbf{B}]. \tag{3.10}$$

The only difference between operators (3.3) and (3.10) is that of a reversal in sign of the background flow term ($\mathbf{v}$ flips sign). Thus the following term may be rearranged such that

$$\int_{\odot} d\mathbf{x} \int d\omega\, \boldsymbol{\lambda} \cdot \mathscr{L}\,\delta\boldsymbol{\xi} = \int_{\odot} d\mathbf{x} \int d\omega\, \delta\boldsymbol{\xi} \cdot \mathscr{L}^{\dagger}\boldsymbol{\lambda}, \tag{3.11}$$

where $\mathscr{L}^{\dagger}$, the adjoint (or dual) operator, acts on Lagrange multiplier $\boldsymbol{\lambda}$ and $\delta\boldsymbol{\xi}$ has been effectively freed. Now, we choose $\boldsymbol{\lambda}$ so as to satisfy the differential equation

$$\mathscr{L}^{\dagger}\boldsymbol{\lambda} - \mathbf{f}^{\dagger} = \mathbf{0}, \tag{3.12}$$

leaving an elegant and simple connection between the variation in misfit and model parameters:

$$\delta\mathscr{I} = -\int_{\odot} d\mathbf{x} \int d\omega\, \boldsymbol{\lambda} \cdot \delta\mathscr{L}\boldsymbol{\xi}. \tag{3.13}$$

Since $\mathscr{L}$ depends solely on background properties (denoted collectively as $\{\beta_s\}$), variations in the operator may be represented as effective functions of $\delta\beta_s$

$$\boldsymbol{\lambda} \cdot \delta\mathscr{L}\boldsymbol{\xi} = \left(\lambda_i \frac{\partial \mathscr{L}_{ij}}{\partial \beta_s} \xi_j\right) \delta\beta_s, \tag{3.14}$$

where Einstein's summation convention is employed and $\mathscr{L} = \{\mathscr{L}_{ij}\}$ is a second-order tensor. Properties β_s in this theory are regarded as being functions only of space; this allows us to define a kernel as

$$K_s = -\int d\omega \left(\lambda_i \frac{\partial \mathscr{L}_{ij}}{\partial \beta_s} \xi_j\right), \tag{3.15}$$

leading to

$$\delta\mathscr{I} = \int_{\odot} d\mathbf{x} \sum_s K_s \delta\beta_s, \tag{3.16}$$

which tells us how to simultaneously solve the inverse problem for all relevant helioseismic quantities:

$$\delta\mathscr{I} = \int_{\odot} d\mathbf{x} \left(K_{\rho}\, \delta\rho + K_{c^2}\, \delta c^2 + \mathbf{K_v}\cdot\mathbf{v} + \mathbf{K_B}\cdot\delta\mathbf{B}\right). \tag{3.17}$$

Note we have not written out a kernel for pressure since it may be determined by considering variations in the equilibrium equation (3.1). For a more detailed treatment, please see section 3.4, equation (3.67). When performing an iterative inversion, it is evident from equation (3.16) that by choosing $\delta\beta_s = -\varepsilon_s K_s$, where $\varepsilon_s > 0$ is a small constant, we arrive at,

$$\delta\mathscr{I} = -\int_\odot d\mathbf{x} \sum_s \varepsilon_s K_s^2 < 0. \tag{3.18}$$

This is the principle of the steepest descent method. More sophisticated inverse algorithms such as the conjugate-gradient method, which uses previous and current gradients to construct the model update at a given iteration level, may be more relevant. Preconditioning, a technique applied to improve the condition number, may also be implemented. The determination of ε requires a "line search" to determine an optimal value (whereas a crude way is to simply set it to some small value, such as 0.02). Thus, we alter the background state by amounts directly proportional to the Fréchet derivative, i.e., we perform the following updates:

$$c^2 \to c^2 - \varepsilon_c\, K_{c^2},$$
$$\rho \to \rho - \varepsilon_\rho\, K_\rho,$$
$$\mathbf{v} \to \mathbf{v} - \varepsilon_v\, \mathbf{K_v},$$
$$\mathbf{B} \to \mathbf{B} - \varepsilon_B\, \mathbf{K_B}. \tag{3.19}$$

The preceding set of equations describes in generality how to pose the helioseismic inverse problem; translational invariance is a specific case of this formalism.

3.3 Measurement functionals

Thus far, we have very generally described the underpinnings of the adjoint method; from this point on, we focus on the primary measurement in time-distance helioseismology: cross correlations. Since we are interested in determining the gradient of the misfit function based on travel times (Eq. [3.4]), we must both appreciate how travel times are computed and quantify their variation with respect to changes in model parameters. Varying equation (3.4), we have

$$\delta\mathscr{I}' = \frac{1}{2}\sum_{q,q'} \mathscr{N}_{qq'}\,[\Delta\tau_{q'}^{(n)}\delta\tau_q + \Delta\tau_q^{(n)}\delta\tau_{q'}], \tag{3.20}$$

$$\delta\mathscr{I}' = \sum_{q,q'} \frac{1}{2}(\mathscr{N}_{qq'} + \mathscr{N}_{q'q})\Delta\tau_{q'}^{(n)}\delta\tau_q, \tag{3.21}$$

$$= \sum_q b_q^{(n)}\delta\tau_q, \tag{3.22}$$

$$b_q^{(n)} = \sum_{q'} \frac{1}{2}(\mathscr{N}_{qq'} + \mathscr{N}_{q'q})\Delta\tau_{q'}^{(n)}, \tag{3.23}$$

where $\Delta\tau_q^{(n)} = [\tau_q^{(n)} - \tau_q^o]$. We do not place the iteration superscript n over the variation in travel time $\delta\tau_q$ because this term implicitly depends on the background, which evolves with each iteration. We apply the following definition of travel time (appendix A of Gizon and Birch 2002)

$$\delta\tau = \int_0^T dt' \, W_{\alpha\beta}(t') \, \delta\mathscr{C}_{\alpha\beta}(t'), \qquad (3.24)$$

where $W_{\alpha\beta}$ is a weight function and $\delta\mathscr{C}_{\alpha\beta}(t')$ the deviation in the cross correlation, α, β are measurement pixel locations, and T is the length of the temporal window. Following Woodard (1997) and Gizon and Birch (2002), we begin by defining the cross correlation

$$\mathscr{C}_{\alpha\beta}(t) = \frac{1}{T} \int_0^T \phi(\mathbf{x}_\alpha, t') \, \phi(\mathbf{x}_\beta, t + t') \, dt', \qquad (3.25)$$

where $\phi(\mathbf{x}, t)$ is the line-of-sight projected wave velocity measured at spatial point $\mathbf{x}$ at the solar photosphere. Appropriate filters and point-spread-function contributions are assumed to have already been incorporated into the definition of $\phi(\mathbf{x}, t)$. Transformed into temporal Fourier space, this becomes

$$\mathscr{C}_{\alpha\beta} = \frac{1}{T} \phi^*(\mathbf{x}_\alpha, \omega) \phi(\mathbf{x}_\beta, \omega). \qquad (3.26)$$

Let Green's tensor for the system of differential equations, denoted by $\mathbf{G}(\mathbf{x}, \mathbf{x}', \omega)$, satisfy

$$\mathscr{L}\mathbf{G} = \delta(\mathbf{x} - \mathbf{x}') \, \mathbf{I}, \qquad (3.27)$$

where $\mathbf{x}$ is termed the "receiver" and $\mathbf{x}'$, "the source," and $\mathbf{I} = \{\delta_{ij}\}$. Similarly, we define the adjoint Green's tensor via

$$\mathscr{L}^\dagger\mathbf{G}^\dagger = \delta(\mathbf{x} - \mathbf{x}') \, \mathbf{I}. \qquad (3.28)$$

Thus for an arbitrary source distribution $\mathbf{S}(\mathbf{x}', \omega)$, the wavefield in temporal Fourier domain is given by

$$\boldsymbol{\xi}(\mathbf{x}, \omega) = \int_\odot d\mathbf{x}' \, \mathbf{G}(\mathbf{x}, \mathbf{x}', \omega) \cdot \mathbf{S}(\mathbf{x}', \omega), \qquad (3.29)$$

and in time domain,

$$\boldsymbol{\xi}(\mathbf{x}, t) = \int_\odot d\mathbf{x}' \int dt' \, \mathbf{G}(\mathbf{x}, \mathbf{x}', t - t') \cdot \mathbf{S}(\mathbf{x}', t'). \qquad (3.30)$$

Similar relations apply to the adjoint wavefield. In order to reduce notational burden, we discontinue explicitly writing the ω dependence, i.e., only source and receiver locations will be included when stating Green's function. In analyses that follow, we shall repeatedly switch positions of the source and receiver. Green's functions in the case of a switched source-receiver pair satisfy the following reciprocity relation (see appendix A.1)

$$\mathbf{G}^{\dagger}(\mathbf{x}',\mathbf{x}) = \mathbf{G}^{T}(\mathbf{x},\mathbf{x}'). \tag{3.31}$$

Observations are typically highly processed versions of the raw solar vector velocity field, subjected to point spreading and phase-speed filtering, line-of-sight projection, etc. Following Gizon and Birch (2002), we introduce vector $\mathscr{G}_j$ to denote Green's function for the filtered, line-of-sight projected velocity

$$\mathscr{G}_j(\mathbf{x},\mathbf{x}') = \mathscr{F}(\mathbf{x},\omega) * l_i(\mathbf{x}) G_{ij}(\mathbf{x},\mathbf{x}'), \tag{3.32}$$

where the convolution is spatio-temporal, $\hat{\mathbf{l}} = \{l_i(\mathbf{x})\}$ is the unit line-of-sight projection vector, and $\mathscr{F}(\mathbf{x},\omega)$ contains all filter terms and the transformation between displacement and observed wavefield velocity. Applying equation (3.31) to (3.32), we may define the reciprocal filtered Green's function

$$\mathscr{G}_j^{\dagger}(\mathbf{x}',\mathbf{x}) = \mathscr{F}(\mathbf{x},\omega) * l_i(\mathbf{x}) G_{ji}^{\dagger}(\mathbf{x}',\mathbf{x}). \tag{3.33}$$

Note that $\mathscr{G}_j(\mathbf{x},\mathbf{x}') \equiv \mathscr{G}_j^{\dagger}(\mathbf{x}',\mathbf{x})$, since all we do is to replace $G_{ij}(\mathbf{x},\mathbf{x}')$ by its adjoint counterpart $G_{ji}^{\dagger}(\mathbf{x}',\mathbf{x})$, to which it is identically equal.

The cross correlation written in terms of Green's tensors, driven by the source $S_k(\mathbf{x},\omega)$, where k is the direction of the dipolar source, is

$$\mathscr{C}_{\alpha\beta} = \frac{1}{T}\int_{\odot} d\mathbf{x}' \int_{\odot} d\mathbf{x}'' \, \mathscr{G}_i^*(\mathbf{x}_\alpha,\mathbf{x}') \, \mathscr{G}_j(\mathbf{x}_\beta,\mathbf{x}'') \, S_i^*(\mathbf{x}',\omega) \, S_j(\mathbf{x}'',\omega). \tag{3.34}$$

Measured cross correlations are typically averaged over a large number of source-correlation times, allowing us to treat it as an ensemble average over many source realizations. In other words, we consider a limit cross correlation that has detached itself from detailed properties of source action and is sensitive only to the statistical quantity $\langle S_i^*(\mathbf{x}',\omega) \, S_j(\mathbf{x}'',\omega)\rangle$, where the angled brackets denote ensemble averaging (e.g., Woodard 1997; Gizon and Birch 2002; Tromp et al 2010). In order to render this theory computable, we explicitly assume that sources at disparate spatial points are spatially uncorrelated, allowing us to write

$$\langle S_i^*(\mathbf{x}',\omega) \, S_j(\mathbf{x}'',\omega)\rangle = \delta(\mathbf{x}'-\mathbf{x}'') \, \mathscr{P}_{ij}(\mathbf{x}',\omega), \tag{3.35}$$

where $\mathscr{P}_{ij}$ encapsulates the average temporal power spectrum, correlations between different dipole sources, and the spatial distribution of source amplitudes. Thus the limit cross correlation becomes

$$\langle\mathscr{C}_{\alpha\beta}\rangle = \frac{1}{T}\int_{\odot} d\mathbf{x}' \, \mathscr{G}_i^*(\mathbf{x}_\alpha,\mathbf{x}') \, \mathscr{G}_j(\mathbf{x}_\beta,\mathbf{x}') \, \mathscr{P}_{ij}(\mathbf{x}',\omega). \tag{3.36}$$

Consider a variation in the cross correlation

$$\langle\delta\mathscr{C}_{\alpha\beta}\rangle = \frac{1}{T}\int_{\odot} d\mathbf{x}' \, [\mathscr{G}_i^*(\mathbf{x}_\alpha,\mathbf{x}') \, \delta\mathscr{G}_j(\mathbf{x}_\beta,\mathbf{x}') + \delta\mathscr{G}_i^*(\mathbf{x}_\alpha,\mathbf{x}') \, \mathscr{G}_j(\mathbf{x}_\beta,\mathbf{x}')] \, \mathscr{P}_{ij}, \tag{3.37}$$

where we have chosen to neglect changes in properties of the power spectrum, i.e., $\delta \mathscr{P}_{ij}(\mathbf{x}', \omega) = 0$. We invoke the first-Born approximation to describe variations in Green's tensor due to changes in properties of the background medium

$$\mathscr{L} \, \delta \mathbf{G} = -\delta \mathscr{L} \, \mathbf{G}. \tag{3.38}$$

Using Green's identity, we recover the following expression for $\delta G_{ij}(\mathbf{x}, \mathbf{x}')$

$$\delta G_{ij}(\mathbf{x}, \mathbf{x}') = - \int_{\odot} d\mathbf{x}'' \, G_{ik}(\mathbf{x}, \mathbf{x}'') \, [\delta \mathscr{L} \, \mathbf{G}(\mathbf{x}'', \mathbf{x}')]_{kj}, \tag{3.39}$$

where the spatial coordinate in $\delta \mathscr{L}$ is $\mathbf{x}''$. Finally, we have

$$\delta \mathscr{G}_j(\mathbf{x}, \mathbf{x}') = \mathscr{F}(\mathbf{x}, \omega) * [l_i \, \delta G_{ij}] = - \int_{\odot} d\mathbf{x}'' \, \mathscr{G}_k(\mathbf{x}, \mathbf{x}'') \, [\delta \mathscr{L} \, \mathbf{G}(\mathbf{x}'', \mathbf{x}')]_{kj}, \tag{3.40}$$

where $\delta \mathscr{L}$ is a function of $\mathbf{x}''$ and the filter $\mathscr{F}(\mathbf{x}, \omega)$ acts only on $l_i(\mathbf{x}) \, G_{ik}(\mathbf{x}, \mathbf{x}'')$. Considering only the first term in the variation of the cross correlation in equation (3.37), we have

$$\langle \delta \mathscr{C}_{\alpha\beta}^1 \rangle = -\frac{1}{T} \int_{\odot} d\mathbf{x} \int_{\odot} d\mathbf{x}' \, [\mathscr{G}_i^*(\mathbf{x}_\alpha, \mathbf{x}') \, \mathscr{G}_k(\mathbf{x}_\beta, \mathbf{x})] \, [\delta \mathscr{L} \, \mathbf{G}(\mathbf{x}, \mathbf{x}')]_{kj} \, \mathscr{P}_{ij}. \tag{3.41}$$

Rearranging the integration order,

$$\langle \delta \mathscr{C}_{\alpha\beta}^1 \rangle = -\frac{1}{T} \int_{\odot} d\mathbf{x} \, \mathscr{G}_k(\mathbf{x}_\beta, \mathbf{x}) \left\{ \delta \mathscr{L}_{kp} \left[\int_{\odot} d\mathbf{x}' \, G_{pj}(\mathbf{x}, \mathbf{x}') \, (\mathscr{G}_i^*(\mathbf{x}_\alpha, \mathbf{x}') \, \mathscr{P}_{ij}) \right] \right\},$$

$$= -\frac{1}{T} \int_{\odot} d\mathbf{x} \, \mathscr{G}_k^\dagger(\mathbf{x}, \mathbf{x}_\beta) \left\{ \delta \mathscr{L}_{kp} \left[\int_{\odot} d\mathbf{x}' \, G_{pj}(\mathbf{x}, \mathbf{x}') \, \left(\mathscr{G}_i^\dagger(\mathbf{x}', \mathbf{x}_\alpha) \, \mathscr{P}_{ij} \right)^* \right] \right\},$$

because $\mathscr{G}_k(\mathbf{x}_\beta, \mathbf{x}) \equiv \mathscr{G}_k^\dagger(\mathbf{x}, \mathbf{x}_\beta)$ (from Eqs. [3.32] and [3.33]) and $\mathscr{P}_{ij}(\omega)$ is real valued. Recalling equation (3.24), and transforming to the temporal Fourier domain, we obtain

$$\delta \tau = \frac{1}{2\pi} \int d\omega \, W_{\alpha\beta}^*(\omega) \, \delta \mathscr{C}_{\alpha\beta}(\omega). \tag{3.42}$$

Now, substituting equation (3.42) into the expression for the misfit (Eq. [3.5] and Eq. [3.22]), we obtain

$$\delta \mathscr{I}_1 = - \sum_q \frac{1}{2\pi T} \int_{\odot} d\mathbf{x} \int d\omega \, W_{\alpha\beta}^*(\omega) \, b_q^{(n)} \, \mathscr{G}_k^\dagger(\mathbf{x}, \mathbf{x}_\beta) \times$$

$$\left\{ \delta \mathscr{L}_{kp} \left[\int_{\odot} d\mathbf{x}' \, G_{pj}(\mathbf{x}, \mathbf{x}') \, (\mathscr{G}_i(\mathbf{x}_\alpha, \mathbf{x}') \, \mathscr{P}_{ij})^* \right] \right\}_k, \tag{3.43}$$

where some bijective mapping function connects q to the cross correlation points (α, β). We define the adjoint field to be

$$\mathbf{\Phi}^\dagger_{\alpha\beta}(\mathbf{x}) = \mathscr{G}^\dagger(\mathbf{x}, \mathbf{x}_\beta) \, W_{\alpha\beta}^*(\omega) \, b_q^{(n)}, \tag{3.44}$$

where subscript k has been dropped from the right side. It is important to note that observations have been assimilated into the adjoint field at this stage; thus, kernels that emerge will be functions of measurements. A subtlety in implementation arises due to the fact that the filter that takes the raw Green's function to the observable is actually applied on the second spatial index, $\mathbf{x}_\beta$. Here we explicitly specify this term

$$\mathscr{G}_k^\dagger(\mathbf{x},\mathbf{x}_\beta) = [\mathscr{F} * (l_i G_{ki}^\dagger)]|_{(\mathbf{x},\mathbf{x}_\beta)} = \int_\odot d\mathbf{x}' \, G_{ki}^\dagger(\mathbf{x},\mathbf{x}') \left[l_i \, \mathscr{F}(\mathbf{x}_\beta - \mathbf{x}',\omega)\right], \quad (3.45)$$

where we have assumed a laterally invariant filter. We arrive at the following adjoint wavefield

$$\boldsymbol{\Phi}^\dagger_{\alpha\beta}(\mathbf{x}) = \int_\odot d\mathbf{x}' \, \mathbf{G}^\dagger(\mathbf{x},\mathbf{x}') \cdot \mathscr{M}(\mathbf{x}',\omega). \quad (3.46)$$

The time-domain representation of this field is

$$\boldsymbol{\Phi}^\dagger_{\alpha\beta}(\mathbf{x},t) = \int_\odot d\mathbf{x}' \int dt' \, \mathbf{G}^\dagger(\mathbf{x},\mathbf{x}',t-t') \cdot \mathscr{M}(\mathbf{x}',t'), \quad (3.47)$$

where $\mathscr{M}$ is a vector whose components are given by

$$\mathscr{M}_i(\mathbf{x},\omega) = W^*_{\alpha\beta}(\omega) \, b_q^{(n)} \left[l_i \, \mathscr{F}(\mathbf{x}_\beta - \mathbf{x},\omega)\right]. \quad (3.48)$$

The forward field represents correlations of the wavefield between every point in the domain and the observed pixel α, and is calculated in a two-step approach (because of the presence of two Green's functions). First we compute the filtered wavefield response to the temporal spectrum of excitation applied at point α

$$\boldsymbol{\eta}(\mathbf{x},\omega) = \int_\odot d\mathbf{x}' \, \mathbf{G}^\dagger(\mathbf{x},\mathbf{x}') \cdot \mathscr{D}, \quad (3.49)$$

where using equations (3.45) and (3.48), we define the source

$$\mathscr{D}_j(\mathbf{x},\mathbf{x}',\omega) = \mathscr{F}(\mathbf{x}_\alpha - \mathbf{x}',\omega) \, l_i \, \mathscr{P}_{ij}(\mathbf{x},\omega) \quad (3.50)$$

In time domain this equation is

$$\boldsymbol{\eta}(\mathbf{x},t) = \int_\odot d\mathbf{x} \int_0^t dt' \, \mathbf{G}^\dagger(\mathbf{x},\mathbf{x}',t-t') \cdot \mathscr{D}(\mathbf{x},\mathbf{x}',t'). \quad (3.51)$$

This response in reverse time is applied as a source again, leading to the forward wavefield

$$\boldsymbol{\Phi}_\alpha(\mathbf{x}) = \int_\odot d\mathbf{x}' \, \mathbf{G}(\mathbf{x},\mathbf{x}') \cdot \boldsymbol{\eta}^*(\mathbf{x}',\omega), \quad (3.52)$$

whose time-domain representation is given by

$$\boldsymbol{\Phi}_\alpha(\mathbf{x},t) = \int_\odot d\mathbf{x}' \int_0^t dt' \, \mathbf{G}(\mathbf{x},\mathbf{x}',t-t') \cdot \boldsymbol{\eta}(\mathbf{x}',-t'). \quad (3.53)$$

We arrive at the following interaction integral

$$\delta \mathscr{I}_1 = -\sum_{\alpha,\beta} \frac{1}{2\pi T} \int_{\odot} d\mathbf{x} \int d\omega \; \boldsymbol{\Phi}^{\dagger}{}_{\alpha\beta} \cdot (\delta \mathscr{L} \; \boldsymbol{\Phi}_{\alpha}). \tag{3.54}$$

The second contribution to the variation in misfit may be written as

$$\delta \mathscr{I}_2 = -\sum_{q} \frac{1}{2\pi T} \int_{\odot} d\mathbf{x} \int d\omega \; W^{*}_{\alpha\beta}(\omega) \; b^{(n)}_q \; \mathscr{G}_k^{\dagger *}(\mathbf{x},\mathbf{x}_{\alpha}) \times$$
$$\left\{ \delta \mathscr{L}^{*} \left[\int_{\odot} d\mathbf{x}' \; G^{*}_{pj}(\mathbf{x},\mathbf{x}') \; (\mathscr{G}_i(\mathbf{x}_{\beta},\mathbf{x}') \; \mathscr{P}_{ij}(\mathbf{x}',\omega)) \right] \right\}_k, \tag{3.55}$$

and since all of these functions have purely real temporal representations, integration over frequency allows us to use the relation $\delta \mathscr{I}_2^{*} = \delta \mathscr{I}_2$, whereby

$$\delta \mathscr{I}_2 = -\sum_{q} \frac{1}{2\pi T} \int_{\odot} d\mathbf{x} \int d\omega \; W_{\alpha\beta}(\omega) \; b^{(n)}_q \; \mathscr{G}_k^{\dagger}(\mathbf{x},\mathbf{x}_{\alpha}) \times$$
$$\left\{ \delta \mathscr{L} \left[\int_{\odot} d\mathbf{x}' \; G_{pj}(\mathbf{x},\mathbf{x}') \; \left(\mathscr{G}_i^{\dagger}(\mathbf{x}',\mathbf{x}_{\beta}) \; \mathscr{P}_{ij}(\mathbf{x}',\omega)\right)^{*} \right] \right\}_k, \tag{3.56}$$

which resembles equation (3.43), except for the adjoint source now being slightly different and with adjoint and source points exchanged. The algorithm for computing this second term remains unchanged from that required for the first contribution. The total misfit variation is given by

$$\delta \mathscr{I} = -\sum_{\alpha,\beta} \frac{1}{2\pi T} \int_{\odot} d\mathbf{x} \int d\omega \; \boldsymbol{\Phi}^{\dagger}{}_{\alpha\beta} \cdot (\delta \mathscr{L} \; \boldsymbol{\Phi}_{\alpha}) + \boldsymbol{\Phi}^{\dagger}{}_{\beta\alpha} \cdot (\delta \mathscr{L} \; \boldsymbol{\Phi}_{\beta}), \tag{3.57}$$

where wavefields and corresponding sources are read off from the two misfit contributions, $\delta \mathscr{I}_1$ (Eq. [3.43]) and $\delta \mathscr{I}_2$ (Eq. [3.56]). In summary, we have deconstructed the meaning of the quantity "travel time," and expressed it in terms of primitive wavefield descriptors such as Green's functions and sources. Next, we studied its variation with respect to small perturbations to the wave operator - the first step in determining the Fréchet derivative. Having quantified its variation, we decomposed the Fréchet derivative into two constituent wavefields, whose convolution, mediated by an operator, reduces to the sensitivity kernel for that parameter.

3.4 Computing Sensitivity Kernels

With suitable notation and mathematics in place, we now describe convolution relations between forward and adjoint wavefields which give sensitivity kernels for various model parameters, such as background flows, sound speed, density, and magnetic fields. The latter two, in addition to the equilibrium equation, determine the corresponding variation in pressure. We begin with flow kernels; changes in isolation to the flow operator are written as $\delta \mathscr{L} = -2i\omega\rho\mathbf{v}\cdot\nabla$. Substituting this into equation (3.54), we obtain

$$\delta \mathscr{I}_1 = 2i \sum_{\alpha,\beta} \frac{1}{2\pi T} \int_\odot d\mathbf{x} \int d\omega\, \omega\rho\, \mathbf{\Phi}^\dagger{}_{\alpha\beta} \cdot (\mathbf{v}\cdot\nabla)\mathbf{\Phi}_\alpha$$

$$= \int_\odot d\mathbf{x}\, \mathbf{v}\cdot\mathbf{K}_\mathbf{v}^{(1)}, \tag{3.58}$$

where,

$$\mathbf{K}_\mathbf{v}^{(1)}(\mathbf{x}) = 2i\rho \sum_{\alpha,\beta} \frac{1}{2\pi T} \int d\omega\, \omega\, (\nabla\mathbf{\Phi}_\alpha)\cdot\mathbf{\Phi}^\dagger{}_{\alpha\beta}. \tag{3.59}$$

Alternately, written in time domain, the flow sensitivity kernel becomes

$$\mathbf{K}_\mathbf{v}^{(1)}(\mathbf{x}) = -2 \sum_{\alpha,\beta} \frac{1}{T} \int dt\, \rho[\nabla\partial_t\mathbf{\Phi}_\alpha(t)]\cdot\mathbf{\Phi}^\dagger{}_{\alpha\beta}(-t), \tag{3.60}$$

where for sake of convenience, we do not explicitly state the $\mathbf{x}$ dependence of the two fields. If we were to compute forward and adjoint fields based on equations (3.52) and (3.44), then interaction (3.60) between forward and time-reversed adjoint fields gives us the desired sensitivity kernel. The second contribution to misfit (and therefore the kernel) must be computed and added to equation (3.60), i.e.,

$$\mathbf{K}_\mathbf{v} = \mathbf{K}_\mathbf{v}^{(1)} + \mathbf{K}_\mathbf{v}^{(2)}, \tag{3.61}$$

where

$$\mathbf{K}_\mathbf{v}^{(2)}(\mathbf{x}) = -2 \sum_{\alpha,\beta} \frac{1}{T} \int dt\, \rho\, (\nabla\partial_t\mathbf{\Phi}_\beta(t))\cdot\mathbf{\Phi}^\dagger{}_{\beta\alpha}(-t). \tag{3.62}$$

Next, we consider perturbations to sound speed, $\delta\mathscr{L} = -\nabla(\rho\delta c^2\,\nabla\cdot)$. Substituting this in equation (3.54), we have

$$\delta\mathscr{I}_1 = \sum_{\alpha,\beta} \frac{1}{2\pi T} \int_\odot d\mathbf{x} \int d\omega\, \mathbf{\Phi}^\dagger{}_{\alpha\beta}\cdot\nabla(\rho\delta c^2\,\nabla\cdot\mathbf{\Phi}_\alpha) \tag{3.63}$$

$$= \sum_{\alpha,\beta} \frac{1}{2\pi T} \int_\odot d\mathbf{x} \int d\omega\, \nabla\cdot(\rho\delta c^2\,\mathbf{\Phi}^\dagger{}_{\alpha\beta}\nabla\cdot\mathbf{\Phi}_\alpha) - \rho\delta c^2\,\nabla\cdot\mathbf{\Phi}^\dagger{}_{\alpha\beta}\,\nabla\cdot\mathbf{\Phi}_\alpha.$$

The first term reduces to a surface integral at the domain boundaries and may therefore be dropped (having assumed homogeneous boundary conditions as $\mathbf{x}\to\infty$). The sound-speed kernel reduces to

$$\delta\mathscr{I}_1 = \int_\odot d\mathbf{x}\, \delta\ln c^2\, K_{c^2}^{(1)}, \tag{3.64}$$

$$K_{c^2}^{(1)}(\mathbf{x}) = -\rho c^2 \sum_{\alpha,\beta} \frac{1}{2\pi T} \int d\omega\, \nabla\cdot\mathbf{\Phi}^\dagger{}_{\alpha\beta}\, \nabla\cdot\mathbf{\Phi}_\alpha. \tag{3.65}$$

Alternately, in time domain, the sound-speed kernel is obtained upon computing

$$K_{c^2}^{(1)}(\mathbf{x}) = -\rho c^2 \sum_{\alpha,\beta} \frac{1}{T} \int dt \, \boldsymbol{\nabla} \cdot \boldsymbol{\Phi}^\dagger_{\alpha\beta}(-t) \, \boldsymbol{\nabla} \cdot \boldsymbol{\Phi}_\alpha(t). \tag{3.66}$$

In order to derive kernel expressions for magnetic field, density, and pressure, which are additionally constrained by the equilibrium equation, we consider small perturbations around (3.1), namely

$$\boldsymbol{\nabla}\delta p = \mathbf{g}\,\delta\rho + \rho\,\delta\mathbf{g} + (\boldsymbol{\nabla} \times \delta\mathbf{B}) \times \mathbf{B} + (\boldsymbol{\nabla} \times \mathbf{B}) \times \delta\mathbf{B}. \tag{3.67}$$

We may ignore perturbations to the gravitational field arising from surface phenomena, such as sunspots or flows in the convection zone, because an overwhelming fraction of solar mass is concentrated within the radiative interior. We have

$$\delta \mathscr{I}_1 = \frac{1}{2\pi T} \int_\odot d\mathbf{x} \, \boldsymbol{\Phi}^\dagger_{\alpha\beta} \cdot \boldsymbol{\nabla}(\boldsymbol{\Phi}_\alpha \cdot \boldsymbol{\nabla}\delta p), \tag{3.68}$$

$$= -\frac{1}{2\pi T} \int_\odot d\mathbf{x} \, \boldsymbol{\nabla} \cdot \boldsymbol{\Phi}^\dagger_{\alpha\beta} \boldsymbol{\Phi}_\alpha \cdot \boldsymbol{\nabla}\delta p, \tag{3.69}$$

$$= -\frac{1}{2\pi T} \int_\odot d\mathbf{x} \, \boldsymbol{\nabla} \cdot \boldsymbol{\Phi}^\dagger_{\alpha\beta} \boldsymbol{\Phi}_\alpha \cdot [\delta\rho\,\mathbf{g} + (\boldsymbol{\nabla} \times \delta\mathbf{B}) \times \mathbf{B} + (\boldsymbol{\nabla} \times \mathbf{B}) \times \delta\mathbf{B}].$$

Terms involving density and magnetic field in the misfit expression for pressure (Eq. [3.70]) are absorbed into kernel expressions for the former two quantities, respectively. The density kernel follows

$$\delta \mathscr{I}_1 = \int_\odot d\mathbf{x} \, K_\rho^{\prime(1)} \, \delta\rho \tag{3.70}$$

$$K_\rho^{\prime(1)} = -\sum_{\alpha,\beta} \frac{1}{2\pi T} \int d\omega \left[-\omega^2 \, \boldsymbol{\Phi}^\dagger_{\alpha\beta} \cdot \boldsymbol{\Phi}_\alpha - i\omega\Gamma \, \boldsymbol{\Phi}^\dagger_{\alpha\beta} \cdot \boldsymbol{\Phi}_\alpha + c^2 \boldsymbol{\nabla} \cdot \boldsymbol{\Phi}^\dagger_{\alpha\beta} \boldsymbol{\nabla} \cdot \boldsymbol{\Phi}_\alpha \right.$$
$$\left. - \boldsymbol{\Phi}_\alpha \cdot \boldsymbol{\nabla}\mathbf{g} \cdot \boldsymbol{\Phi}^\dagger_{\alpha\beta} - \mathbf{g} \cdot (\boldsymbol{\Phi}_\alpha \cdot \boldsymbol{\nabla}\boldsymbol{\Phi}^\dagger_{\alpha\beta} + \boldsymbol{\Phi}_\alpha \boldsymbol{\nabla} \cdot \boldsymbol{\Phi}^\dagger_{\alpha\beta}) \right], \tag{3.71}$$

which in time domain is

$$K_\rho^{\prime(1)} = -\sum_{\alpha,\beta} \frac{1}{T} \int dt \left\{ \boldsymbol{\Phi}^\dagger_{\alpha\beta}(-t) \cdot \partial_t^2 \boldsymbol{\Phi}_\alpha(t) + \boldsymbol{\Phi}^\dagger_{\alpha\beta}(-t) \cdot [\Gamma * \partial_t \boldsymbol{\Phi}_\alpha(t)] + \right.$$
$$c^2 \boldsymbol{\nabla} \cdot \boldsymbol{\Phi}^\dagger_{\alpha\beta}(-t) \boldsymbol{\nabla} \cdot \boldsymbol{\Phi}_\alpha(t) - \boldsymbol{\Phi}_\alpha(t) \cdot \boldsymbol{\nabla}\mathbf{g} \cdot \boldsymbol{\Phi}^\dagger_{\alpha\beta}(-t)$$
$$\left. - \mathbf{g} \cdot [\boldsymbol{\Phi}_\alpha(t) \cdot \boldsymbol{\nabla}\boldsymbol{\Phi}^\dagger_{\alpha\beta}(-t) + \boldsymbol{\Phi}_\alpha(t)\boldsymbol{\nabla} \cdot \boldsymbol{\Phi}^\dagger_{\alpha\beta}(-t)] \right\} \tag{3.72}$$

This is the same as the kernel expression in, e.g., equation (55) of Liu and Tromp (2008) with their rotation terms and gravitational potential variations (their ψ, $\delta\phi$, and δg_0, respectively) set to zero. Zhu et al (2009), who term this the *impedance kernel* (K_ρ'), showed that it is primarily sensitive to reflection zones and discontinuities. It therefore remains to be seen whether much information about density variations in the solar interior may be extracted from the wavefield.

Dropping the assumption of translation invariance allows us to derive simple vector expressions for magnetic field kernels (see appendix A.2). The effect of small deviations from a background field on travel times is given by the following expression

$$\delta \mathscr{I}_1 = \int_\odot d\mathbf{x} \, \mathbf{K}_\mathbf{B}^{(1)} \cdot \delta \mathbf{B}, \tag{3.73}$$

$$\mathbf{K}_\mathbf{B}^{(1)} = \sum_{\alpha,\beta} \frac{1}{2\pi T} \int d\omega \, \boldsymbol{\nabla} \times [\boldsymbol{\nabla} \times (\boldsymbol{\Phi}_\alpha \times \mathbf{B}) \times \boldsymbol{\Phi}^\dagger{}_{\alpha\beta}] + \boldsymbol{\nabla} \cdot \boldsymbol{\Phi}^\dagger{}_{\alpha\beta} \boldsymbol{\Phi}_\alpha \times [\boldsymbol{\nabla} \times \mathbf{B}]$$

$$+ \left\{ \boldsymbol{\nabla} \times [\boldsymbol{\Phi}^\dagger{}_{\alpha\beta} \times (\boldsymbol{\nabla} \times \mathbf{B})] \right\} \times \boldsymbol{\Phi}_\alpha + \boldsymbol{\Phi}^\dagger{}_{\alpha\beta} \times \left\{ \boldsymbol{\nabla} \times [\boldsymbol{\nabla} \times (\boldsymbol{\Phi}_\alpha \times \mathbf{B})] \right\}$$

$$+ \boldsymbol{\Phi}_\alpha \times \left\{ \boldsymbol{\nabla} \times [\boldsymbol{\nabla} \times (\boldsymbol{\Phi}^\dagger{}_{\alpha\beta} \times \mathbf{B})] \right\} + \boldsymbol{\nabla} \times [\mathbf{B} \times (\boldsymbol{\Phi}_\alpha \boldsymbol{\nabla} \cdot \boldsymbol{\Phi}^\dagger{}_{\alpha\beta})], \tag{3.74}$$

which in time domain is

$$\mathbf{K}_\mathbf{B}^{(1)} = \sum_{\alpha,\beta} \frac{1}{T} \int dt \, \boldsymbol{\nabla} \times [\boldsymbol{\nabla} \times (\boldsymbol{\Phi}_\alpha(t) \times \mathbf{B}) \times \boldsymbol{\Phi}^\dagger{}_{\alpha\beta}(-t)]$$

$$+ \left\{ \boldsymbol{\nabla} \times [\boldsymbol{\Phi}^\dagger{}_{\alpha\beta}(-t) \times (\boldsymbol{\nabla} \times \mathbf{B})] \right\} \times \boldsymbol{\Phi}_\alpha(t)$$

$$+ \boldsymbol{\Phi}^\dagger{}_{\alpha\beta}(-t) \times \left\{ \boldsymbol{\nabla} \times [\boldsymbol{\nabla} \times (\boldsymbol{\Phi}_\alpha(t) \times \mathbf{B})] \right\} \tag{3.75}$$

$$+ \boldsymbol{\Phi}_\alpha(t) \times \left\{ \boldsymbol{\nabla} \times [\boldsymbol{\nabla} \times (\boldsymbol{\Phi}^\dagger{}_{\alpha\beta}(-t) \times \mathbf{B})] \right\}$$

$$+ \boldsymbol{\nabla} \times [\mathbf{B} \times (\boldsymbol{\Phi}_\alpha(t) \boldsymbol{\nabla} \cdot \boldsymbol{\Phi}^\dagger{}_{\alpha\beta}(-t))] + \boldsymbol{\nabla} \cdot \boldsymbol{\Phi}^\dagger{}_{\alpha\beta}(-t) \boldsymbol{\Phi}_\alpha(t) \times [\boldsymbol{\nabla} \times \mathbf{B}].$$

For inversions constrained by equilibrium equation (3.1), we have arrived at kernels for four independent model parameters, namely: sound speed, velocity, density, and magnetic field.

What is the connection between kernels derived here and those computed by, e.g., Birch et al (2004)? Translation invariance implies that everywhere in the domain of interest, differences between predicted and measured travel times are small and that perturbations to the background state are weak. Dividing out the $\Delta \tau_q$ term (setting it to some constant value) from expressions for the kernel and misfit, we arrive at the classical linear helioseismic forward problem

$$\delta \tau = \int_\odot d\mathbf{x} \, \mathbf{K}(\mathbf{x}) \cdot \delta \mathbf{q}(\mathbf{x}), \tag{3.76}$$

where $\delta \mathbf{q}$ is a perturbation of interest.

3.5 Flow and sound-speed kernels

We use the SPARC code, discussed in Section 2.5. A domain of size $250 \times 250 \times 35 \, \text{Mm}^3$ is chosen, where the first two dimensions are horizontal and the third depth. The box straddles the photosphere, extending from 34 Mm below to 1 Mm above.

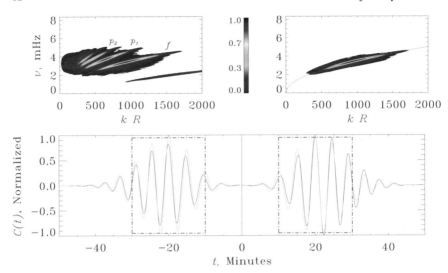

Fig. 3.1 Normalized pre- and post-filtered power spectra of the z component of η on the upper panels and the f-mode dispersion relation $\omega = \sqrt{gk}$ overplotted (dark line; right). Below is shown the normalized predicted f-mode limit cross correlation $\mathscr{C}_{\alpha\beta}(t)$ of the wavefield measured at a pair of points with separation distance $|\alpha - \beta| = 10$ Mm. Overplotted with the thin line is the "exact" cross correlation, estimated by inverse Fourier transforming the power spectrum (Eq. [A.28]). The amplitudes of the negative and positive branches are slightly different but their phases match well. The dot-dash boxes around branches of the limit cross correlation denote the chosen temporal window ($f(t)$ in Eq. [A.21]).

The grid consists of $384 \times 384 \times 300$ points, ensuring a horizontal resolution of 660 km. Vertical grid spacing decreases smoothly from about 250 km at the bottom of the box to around 27 km at the photosphere and above, so designed as to maintain constant acoustic travel time between adjacent pairs of points.

In Figure 3.1, power spectra of pre- and post-filtered intermediate wavefields (vertical component of $\boldsymbol{\eta}$) are shown; we isolate the f-mode for this calculation. The predicted limit cross correlation, $\mathscr{C}_{\alpha\beta}(t)$, obtained by filtering the forward wavefield and extracting the time series at the receiver is also shown. The positive and negative branches differ slightly in amplitude but show good phase agreement.

We display actual wavefields and demonstrate the process of computing kernels graphically in Figure 3.2. The first column shows snapshots of the intermediate wavefield $\boldsymbol{\eta}$ forced by a source at α at three time instants — this wavefield is filtered, time reversed, and fed into the code as a source for the forward wavefield, $\boldsymbol{\Phi}_\alpha$, seen in the second column. The adjoint source is computed using the predicted limit cross correlation that is derived from the forward wavefield and used to drive the adjoint wavefield $\boldsymbol{\Phi}^\dagger_{\alpha\beta}$ (where β is the receiver), depicted in reverse time in the third column at a number of instants. The final two columns show the interaction integral and stages in the construction of the partial kernel. This entire process must be repeated with β as source and α the receiver and its contribution must be added to the partial kernel obtained previously (shown in Figure 3.3).

Cuts through kernels are shown in Figure 3.3. The upper three panels show partial contributions and full kernels, at a depth $z = -0.5$ Mm; vertical cuts through

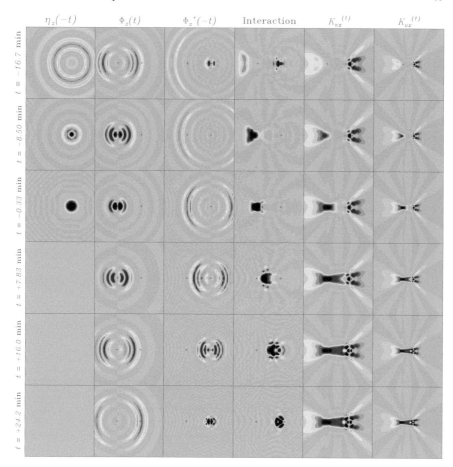

Fig. 3.2 Snapshots in time of various wavefields, interaction integral, and kernel zoomed in and out (akin to Figures 2 and 3 of Tromp et al 2010). The first three columns show vertical components of η, $\boldsymbol{\Phi}_\alpha$, $\boldsymbol{\Phi}^\dagger_{\alpha\beta}$, and the next two display the interaction integral and horizontal flow kernel, $\mathscr{I}_1, K^{(1)}_{v_x}$ with the final column showing a zoomed out picture of the kernel. The adjoint field $\boldsymbol{\Phi}^\dagger_{\alpha\beta}$ and spectral response wavefield η are shown in reverse time in order to highlight the computational algorithm: (1) we time reverse η and feed into the forward calculation and (2) the kernel is calculated via a convolution between the forward and adjoint. The two marks denote locations of source (left) and receiver. Total solar time of the spectral response calculation ($\boldsymbol{\eta}$; left column) is 2.5 hrs, and forward and adjoint are 5 hrs each. Note that the forward field (second column) is centered around the source point while the adjoint (third column) is centered around the receiver.

the $y = 0$ center line for K_{v_x} and K_{v_z} are displayed on the fourth panel. Only the f-mode contributes to the kernel as evidenced by the constancy in sign of the kernel as a function of depth. The x- and y-(anti-) symmetries of kernels are as expected (see, e.g., Birch and Gizon 2007). The two faint horizontal lines seen at $z = -0.2$ Mm and $z = 0.2$ Mm in the vertical cuts of the x- and z-kernels correspond to the excitation depth and observation height, respectively; the intermediate wavefield is

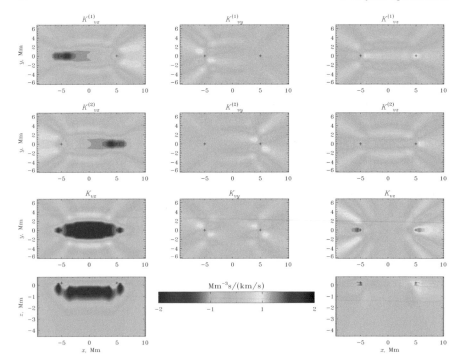

Fig. 3.3 Partial contributions to flow kernels (first two rows) and their sum (third row), displayed at a depth of $z = -0.54$ Mm. A vertical cut through the K_{v_x} and K_{v_z} kernels along the $y = 0$ center line is shown on the fourth row. These are computed around the translationally invariant polytropic background described in Hanasoge et al (2008). Symmetries and magnitudes of the kernels are in line with expectation (Birch and Gizon 2007). Note that there is an extra factor of time in the dimension of the flow kernels that arises from assimilating observed travel times into the kernels.

computed with a source at the former depth and the cross correlation and adjoint sources are injected at the latter height. The integral of the x-flow kernel may be directly estimated from the power spectrum — the two values agree to within a few percent (see appendix A.4).

We also compute the sound-speed kernel for the mean travel time measured using the p_1 ridge. Mean travel times are measured according to equation (A.23). The intermediate, forward, and adjoint simulations are performed and the interaction of the latter two is computed in accordance with equation (3.66). The filtered power spectrum and limit cross correlation for the measurement are shown in Figure 3.4. The positive and negative branches are slightly phase shifted, suggesting the requirement of a larger computational domain. The kernel for this measurement is shown in Figure 3.5. The raypath corresponding to 10 Mm angular (horizontal) distance is also plotted for reference; note similarities to sound-speed kernels computed by Birch et al (2004).

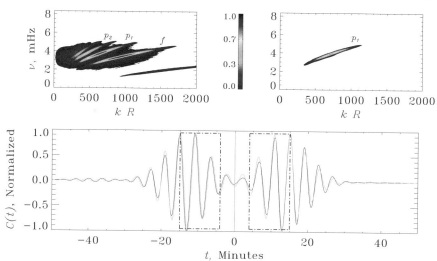

Fig. 3.4 Filtered p_1 spectrum and limit cross correlation for a pair of antennae 10 Mm apart. The thin line depicts the expected cross correlation, estimated from Fourier transforming the power spectrum (Eq. [A.28]).

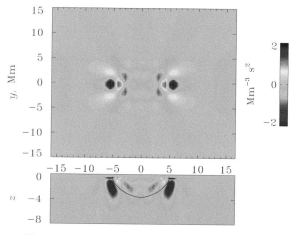

Fig. 3.5 Sound-speed kernel for a mean travel-time measurement using a p_1-mode. The antennae are separated by 10 Mm. The solid line denotes the ray path. A double-bounce p wave, whose ray travel time is approximately 15 minutes (and therefore likely in the temporal window), is also noticeable.

3.6 Magnetic Kernels

A critical aspect to setting up an inverse problem is in appreciating the physical variables to which waves are sensitive. The expressions derived in Section 3.4 tell us that kernels for sound speed and flows are weighted by the density of the model, in contrast to kernels for the primitive magnetic field. Further, variables c and $\boldsymbol{v}$ are forms of wavespeed, which points to the use of Alfvén velocity, $\mathbf{a} = \mathbf{B}/\sqrt{4\pi\rho}$ instead of the primitive $\mathbf{B}$ field. One may conceive of it as a descriptor of the anisotropic wave velocity to which waves are directly sensitive. Straightforward manipulation allows us to rewrite the kernels as follows

$$\delta\mathbf{B} = \delta(\mathbf{a}\sqrt{4\pi\rho}) = \sqrt{4\pi\rho}\,\delta\mathbf{a} + \frac{1}{2}\mathbf{a}\sqrt{4\pi\rho}\,\delta\ln\rho, \qquad (3.77)$$

which gives

$$\sqrt{4\pi\rho}\,\mathbf{K_B} = \mathbf{K_a} \qquad K'_\rho \rightarrow K'_\rho + \frac{1}{2}\mathbf{K_a}\cdot\mathbf{a}, \qquad (3.78)$$

thus providing a new expression for variations in the misfit

$$\delta\chi = -\int_\odot d\mathbf{x}\,K_c\,\delta c + \mathbf{K}_v\cdot\delta v + \mathbf{K_a}\cdot\delta\mathbf{a} + K'_\rho\,\delta\ln\rho. \qquad (3.79)$$

We note that the first three terms represent three types of wavespeeds, an isotropic sound speed, an advection-related flow velocity, and lastly, an intrinsically anisotropic velocity. Although not shown here, weighting the magnetic field kernels by the square-root of density redistributes incoherent sensitivity from the upper-most atmospheric layers to the photosphere and shallow interior. The transformation for the density kernel in equation (3.77) now contains a contribution from the Alfvén velocity, and could in principle be used to image reflections off sharp velocity contrasts.

The single-scattering first-Born approximation cannot capture the full scope of wave propagation in strong perturbations such as sunspots (i.e., with respect to the quiet Sun; e.g., Gizon et al 2006). This implies that inversions for the sub-surface structure of sunspots are likely to require an iterative algorithm, since we have to sequentially refine the predicted travel times, which are nonlinearly related to changes in the model. Thus, in the analysis here, we construct a 'sunspot' in MHS equilibrium (Eq. [3.1]) and determine sensitivity kernels relative to this model.

We introduce a 2-D stream function $\psi(x,z)$ such that the magnetic field is given by $\mathbf{B} = (-\partial_z\psi, \partial_x\psi)$. Since $\mathbf{g} = (0, -g)$, Lorentz forces in the x direction are solely balanced by the pressure gradient in equation (3.1), i.e., $\partial_x p = \partial_x(B_x^2/2 - B_z^2/2) + \partial_z(B_x B_z)$. From this equation we calculate the pressure distribution required to support this field configuration and then use the z component of equation (3.1) to obtain the associated density. Generating an MHS state is non-trivial since density and pressure decrease exponentially as a function of height above the photosphere; consequently, a large range of choices for the field configuration results in negative pressures or densities or both. Field configurations with strong horizontal and vertical fields also require the action of flows to maintain force balance, an aspect we do not consider here because the complexity of such a model renders difficult the interpretation of the attendant kernels. We show one example field configuration in Figure 3.6.

A major difficulty in simulating wave propagation through strong magnetic fields is that (also see Section 2.6.3) Alfvén speed $\|\mathbf{a}\|$ becomes extremely large in the atmospheric layers of the Sun (due to the exponentially rapidly decreasing density), resulting in a very stiff differential equation. Further, wave travel times are very weakly sensitive to the dynamics of these layers because the modes are trapped below the photosphere. A multiplicative prefactor is introduced to control the amplitude of the Lorentz force terms in (3.3), e.g., Cameron et al (2008); Rempel et al (2009). However, this method results in a model that is not seismically reciprocal (e.g., Hanasoge et al 2011), a central requirement in the formal interpretation of helioseismic measurements and the determination of sensitivity kernels. Here, in order to maintain seismic reciprocity while still saturating the Alfvén speed at 40 km/s, we directly multiply the magnetic field by a prefactor. While this results in a background field configuration that has a non-zero divergence, we note that small-amplitude oscillations about this field are still divergence free. Further, in the scheme of linear inversions for magnetic structure, the divergence-free nature of the background field is not a strict requirement but could be considered a regularization term. We perform linear magneto-hydrodynamic (MHD) wave propagation simulations in Cartesian geometry, using the pseudo-spectral code SPARC (Section 2.5). Because we restrict ourselves to a 2-D field configuration in this problem, Aflvén waves are disallowed and only magneto-acoustic fast and slow waves propagate.

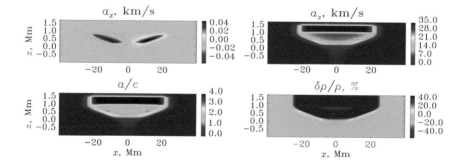

Fig. 3.6 Magnetic field configuration in our calculations. Top panels show Aflvén speeds $a_x = B_x/\sqrt{4\pi\rho}$ and $a_z = B_z/\sqrt{4\pi\rho}$, which are signed quantities. The bottom left panel is the ratio of the absolute Alfvén speed to the local sound speed and is seen to be on the order of 1 at the photosphere. The field is relatively weak with the highest Alfvén speed around 35 km/s and a Wilson depression of 250 km. (for an expanded view, see Figure 1 of the supplemental material of Hanasoge et al 2012a)

We focus here on the diagnostic ability of the surface f and acoustic p_1-modes, so chosen because of their significant sensitivity to surface layers. The measurement consists of ridge filters applied to isolate these modes. The sunspot is assumed to be located at disk center, implying that the line-of-sight component is co-aligned with the (vertical) z axis. Thus the vertical wavefield displacement is used to define the cross correlation measurement. We show the power spectra and cross correlations in Figure 3.7. We employ the linear travel-time definition (Gizon and Birch 2002, 2004).

Figure 3.8 (see also Figures 5 and 6 in the supplemental material of Hanasoge et al 2012a) displays the sensitivity of the surface f-mode to the sunspot. Because we model waves as finite spatial objects, their sensitivities extend beyond just the ray path. It can be seen that the effect of the spot is significant in that the kernels are noticeably asymmetric between the point pair. The time shifts induced by the magnetic field are considerable, comparable in magnitude to those induced by flow and thermal perturbations. There are hints of mode conversion from f to p_1 in the difference kernel for sound speed (top), just below the pixel on the right.

In Figure 3.9 (see also Figures 7 and 8 in the supplemental material of Hanasoge et al 2012a), we show a set of difference p_1-mode kernels for a point pair separated by a distance of 25 Mm, respectively. Because the magnetic field is relatively weak compared to a sunspot, the acoustic p_1-mode, whose energy is focused in the subsurface layers, is much less affected by the field than the f-mode. Symmetry is nearly completely restored to the p_1 kernels.

The Alfvén speed kernels for both f- and p_1-modes show features of high spatial frequency, and contain signatures of fast and slow magneto-acoustic waves. In the umbral regions of the sunspot, waves of high spatial frequency are seen to be propagating towards the interior (plausibly slow waves).

3.7 Performing Inversions, Computational Algorithm, & Cost

In this section, we describe how the adjoint technique may be applied efficiently to perform large-scale inversions using helioseismic data. We begin with the concept of an *event kernel*, discussed in, e.g., Bamberger et al (1982); Igel et al (1996); Tromp et al (2005); Tape et al (2009), the focus of the inverse procedure. Consider a seismic event (i.e., a source at) α whose signature is recorded at some N locations. In a computational sense, the predicted wavefield generated by this source event is encoded in the forward wavefield, while observations are assimilated into the adjoint wavefield. The expense involved in computing event kernels scales linearly with number of sources α, independently of the number of observation locations. This is because (as will be shown here) all N observations may simultaneously be injected at corresponding station locations to produce the adjoint field. In the translationally invariant case, the event kernel may be obtained by summing up N appropriately rotated and translated kernels, each weighted by the relevant travel-time shift

$$K_\alpha(\mathbf{x}) = \sum_{\beta=1}^{N} \Delta \tau_{\alpha\beta} K_{\alpha\beta}(\mathbf{x}). \qquad (3.80)$$

We formulate algorithmic details associated with incorporating large numbers of observations into the inversion (see also Tromp et al 2010). Let us choose M master

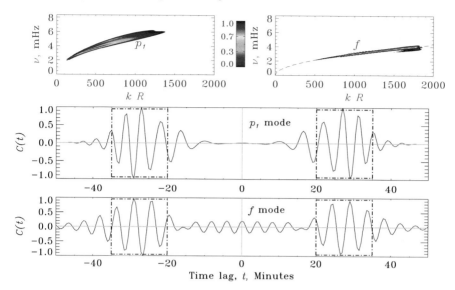

Fig. 3.7 Expectation value of the power spectrum of the p_1 and f ridge-filtered measurements (top panels). The limit cross correlation $\mathscr{C}(t)$ between a point 15 Mm from the left of the sunspot center to a point 10 Mm on the right of the center is shown for the p_1 measurement (middle panel). The f-mode cross correlation is between the center of the sunspot and a point 10 Mm to the right (bottom panel). See Figures 3.8 and 3.9 also. The positive-time branch is sensitive to waves that first arrive at one measurement point and subsequently at the other and vice versa. The loss of translational variance implies that the absolute locations of the points matter. The dot-dash boxes indicate the measurement windows. Travel-time shifts of waves are obtained by estimating the deviation of the cross correlation from a reference wavelet. *Mean* travel times, defined as the average of the time shifts of oppositely traveling waves, are thought to be largely sensitive to structure. *Difference* travel times, defined as the difference between the shifts, are considered primarily sensitive to symmetry-breaking flows. (for an expanded view, see Figure 2 of the supplemental material of Hanasoge et al 2012a)

pixels, which are correlated with signals measured at N pixels, i.e., $MN + M(M - 1)/2$ correlations in total, a number that scales as $O(MN)$ since $M \ll N$. The associated misfit may be written as

$$\delta \mathscr{I} = -\sum_{\beta=1}^{N} \sum_{\alpha=1}^{M} \frac{1}{2\pi T} \int_{\odot} d\mathbf{x} \int d\omega \left(\boldsymbol{\Phi}^{\dagger}{}_{\alpha\beta} \cdot \delta \mathscr{L} \boldsymbol{\Phi}_{\alpha} + \boldsymbol{\Phi}^{\dagger}{}_{\beta\alpha} \cdot \delta \mathscr{L} \boldsymbol{\Phi}_{\beta} \right). \quad (3.81)$$

We attempt to moderate computational cost by absorbing the summation over N into forward and adjoint sources, suitably redefined. The first contribution to misfit may be rewritten as

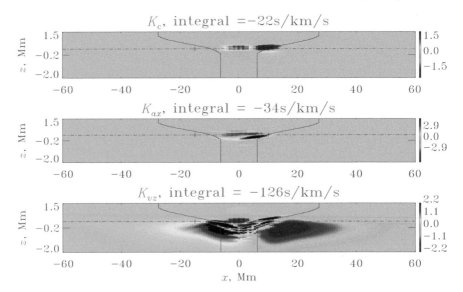

Fig. 3.8 f-mode (surface) wavespeed kernels for a difference travel-time measurement between a point pair 10 Mm apart. Kernels sensitive to isotropic sound speed, Alfvén speed a_x, and vertical flows v_z are shown. The boundary of the spot, marked by the solid black line, is much smaller than the horizontal wavelength. The horizontal dot-dash line denotes the height at which observations are made in the quiet Sun and the symbols mark the measurement points. The f-mode is seen to be significantly affected by the spot, as seen in the loss in symmetry of the kernels. Signatures of magneto-acoustic slow and fast modes and hints of conversion to acoustic p_1 may be plausibly discerned upon examination. The integrals of the kernels show that the travel times are significantly affected by the presence of even this relatively weak magnetic field. (for an expanded view, see Figure 3 of the supplemental material of Hanasoge et al 2012a)

$$\delta \mathscr{I}_1 = -\sum_{\beta=1}^{N} \sum_{\alpha=1}^{M} \frac{1}{2\pi T} \int_{\odot} d\mathbf{x} \int d\omega \, \boldsymbol{\Phi}^{\dagger}{}_{\alpha\beta} \cdot \delta \mathscr{L} \boldsymbol{\Phi}_{\alpha}$$

$$= -\sum_{\alpha=1}^{M} \frac{1}{2\pi T} \int_{\odot} d\mathbf{x} \int d\omega \, \bar{\Phi}_{\alpha}^{\dagger} \cdot \delta \mathscr{L} \boldsymbol{\Phi}_{\alpha}, \qquad (3.82)$$

where the adjoint source and wavefield are given by

$$\mathscr{M}_i(\mathbf{x}) = l_i \sum_{\beta=1}^{N} \mathscr{F}(\mathbf{x}_\beta - \mathbf{x}, \omega) \, W_{\alpha\beta}^* \, b_q^n, \qquad (3.83)$$

$$\bar{\Phi}_{\alpha}^{\dagger}(\mathbf{x}) = \int_{\odot} d\mathbf{x}' \, \mathbf{G}(\mathbf{x}, \mathbf{x}') \cdot \mathscr{M}(\mathbf{x}', \omega), \qquad (3.84)$$

and the forward wavefield is as stated in equation (3.52). All N cross correlations of slave pixels with master pixel α are subsumed into one adjoint calculation. Computationally, this is accomplished by constructing adjoint source (3.83) as a sum over all slave pixels that are correlated with α. The partial event kernel $K_\alpha^{(1)}$ may be

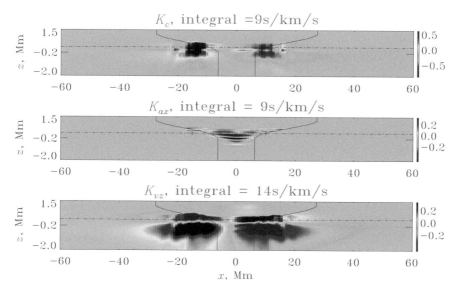

Fig. 3.9 p_1-mode wavespeed kernels for a difference travel-time measurement between a point pair 25 Mm apart. The panels from top to bottom show kernels sensitive to sound speed (top), Alfvén speeds a_x, and vertical flows v_z. The boundary of the spot, marked by the solid black line, is much smaller than the horizontal wavelength. The horizontal dot-dash line denotes the height at which observations are made in the quiet Sun and the symbols mark the measurement points. Plausible signatures of slow modes propagating down into the tube may be discerned in the middle panel. (for an expanded view, see Figure 4 of the supplemental material of Hanasoge et al 2012a)

computed using the wavefields in equation (3.82) together with kernel expressions stated in the preceding section. The second contribution requires some manipulation and redefinition, namely

$$\delta \mathscr{I}_2 = - \sum_{\beta=1}^{N} \sum_{\alpha=1}^{M} \frac{1}{2\pi T} \int_{\odot} d\mathbf{x} \int d\omega \, \mathbf{\Phi}^{\dagger}_{\beta\alpha} \cdot \delta \mathscr{L} \mathbf{\Phi}_{\beta}$$

$$= - \sum_{\alpha=1}^{M} \frac{1}{2\pi T} \int_{\odot} d\mathbf{x} \int d\omega \, \bar{\Phi}^{\dagger}_{\alpha} \cdot \delta \mathscr{L} \bar{\Phi}_{\alpha}, \tag{3.85}$$

$$\mathscr{M}_i(\mathbf{x}) = l_i \mathscr{F}(\mathbf{x}_{\alpha} - \mathbf{x}, \omega), \tag{3.86}$$

$$\bar{\Phi}^{\dagger}_{\alpha}(\mathbf{x}) = \int_{\odot} d\mathbf{x}' \, \mathbf{G}(\mathbf{x}, \mathbf{x}') \cdot \mathscr{M}(\mathbf{x}', \omega), \tag{3.87}$$

$$\mathscr{D}_{\alpha}(\mathbf{x}, \mathbf{x}', \omega) = \sum_{\beta=1}^{N} W_{\alpha\beta} b_q^{(n)} \, \mathscr{F}(\mathbf{x}' - \mathbf{x}_{\beta}) \hat{\mathbf{I}} \cdot \mathscr{P}(\mathbf{x}, \omega), \tag{3.88}$$

$$\bar{\boldsymbol{\eta}}_{\alpha}(\mathbf{x}, \omega) = \int_{\odot} d\mathbf{x}' \, \mathbf{G}^{\dagger}(\mathbf{x}, \mathbf{x}') \cdot \mathscr{D}_{\alpha}(\mathbf{x}, \mathbf{x}', \omega), \tag{3.89}$$

$$\bar{\Phi}_{\alpha} = \int_{\odot} d\mathbf{x}' \, \mathbf{G}(\mathbf{x}, \mathbf{x}') \cdot \bar{\boldsymbol{\eta}}_{\alpha}(\mathbf{x}', \omega). \tag{3.90}$$

The second contribution is constructed by interacting the two wavefields according to equation (3.85) and added to $K_\alpha^{(1)}$ to complete the calculation of the full event kernel. Thus the vast number of observations of the solar wavefield may all be assimilated into the inversion but with a finite $O(M)$ number of calculations. The M master pixels may be chosen to ensure the greatest coverage within the region of interest, whose locations could be decided by criteria such as maximizing the sum of distances between point pairs. The algorithm, depending on whether sensitivity kernels are being computed or inversions are performed may be stated in the following manner:

- *Master Pixels*: Choose a set of M master (α) and N slave (β) pixels with $M \ll N$. For instance, a constellation of points surrounding a sunspot or active region.
- *Intermediate Wavefield* ($\boldsymbol{\eta}, \bar{\boldsymbol{\eta}}$): If the intent is to compute kernels, source (3.50) is applied (Eq. [3.49]) and the resulting wavefield is saved at all points where the wave excitation source is non-zero. Alternately, when performing inversions, two types of sources, given by (3.50) and (3.88), must be applied. Because this wavefield is used to drive the forward simulation, it must be saved at a sufficient number of temporal points. This does not demand large storage requirements since only 2-D slices are written out (for all practical purposes, wave excitation occurs at one depth). The driving source for the calculation of a sensitivity kernel is given by (3.49) and for the event kernel (3.88). This is termed the *generating wavefield* by Tromp et al (2010).
- *Forward Wavefield* ($\boldsymbol{\Phi}, \bar{\boldsymbol{\Phi}}$): Driven by the time-reversed intermediate wavefield displacement ($\boldsymbol{\eta}, \bar{\boldsymbol{\eta}}$) injected at the nominal excitation depth, with the specific choice of sources dependent on whether an event kernel (3.89) or a sensitivity kernel (3.52) is being computed. The 3-D wavefield is saved at a cadence of 30 seconds (Nyquist frequency of 16.66 mHz). This is termed the *ensemble forward wavefield* by Tromp et al (2010).
- *Adjoint Source* ($\mathcal{M}$): The time history of the forward wavefield extracted at the observation height is filtered according to equation (3.32) and time series at all slave pixels are isolated. These form the predicted limit cross correlations for those point pairs. We now determine the adjoint source according to equations (3.48), (3.83), or (3.86) as the case may be (i.e., computing kernels between a point pair or performing an inversion using large numbers of observations). Note this is the stage where observations are assimilated into the inversion.
- *Adjoint Wavefield & Partial Kernels* ($\boldsymbol{\Phi}^\dagger, \bar{\boldsymbol{\Phi}}^\dagger$): The former is evaluated according to equations (3.44), (3.84), or (3.87) as the case may be. The 3-D adjoint wavefield is saved at the same cadence as that of the forward. We may then compute kernels according to interaction integral (3.54). Each sensitivity or event kernel has two contributions which must be added together.
- *Temporal Length & Computational Domain Size*: Simulations must be run for at least as long as it takes for waves to arrive from the farthest contributing source to the observation points. The farther the source is, the greater the effects of damping and geometric spreading and thus the contribution of a source diminishes with distance from observation points.

- *Storage Cadence*: Five to ten points per temporal wavelength is a reasonable rule of thumb. This is done in order to maximize the accuracy in evaluating the interaction integral (3.54) while not placing unnecessary demands on storage. Of course, this step may be obviated if one were to apply the algorithm of Liu and Tromp (2008).
- *Boundary Conditions*: Highly absorbent boundary conditions are recommended in order that waves that have propagated out do not return to the region of interest.

For a fixed resolution and temporal extent of the calculation, both storage and computational expense increase linearly with the number of master pixels, i.e., computations scale as $O(5M)$ Green's function calculations, where M is the number of master pixels. This is because the intermediate wavefield needs only be computed for a time extent $T/2$ whereas the forward and adjoint wavefields must be calculated over a temporal length T. Storage cost scales approximately as $O(4M)$. The power of this technique is twofold, firstly in being able to compute all kernels relevant to a given measurement simultaneously from the adjoint and forward wavefields, and secondly, in assimilating as many observations as desired in order to perform the inversion.

Rapid convergence, i.e., reduction in misfit, is a desirable quality of an inverse technique. Two well-known drawbacks of the steepest descent method are that it converges very slowly for problems where the condition number is large and the convergence rate is very sensitive to the local step-size (ε in Eq. [3.19]). A much more popular and powerful technique is the conjugate-gradient method, which utilizes misfit gradients at current and previous iterations in order to determine the directionality and magnitude of the step to be taken. Preconditioning gradients in order to reduce the condition number and improve convergence characteristics is also a typically employed procedure. We shall not describe these issues in any greater detail at present but merely note their importance and that they need be addressed in any inverse procedure. For an in-depth discussion of these topics, see, e.g., Tape et al (2007).

An important aspect of the outcome of an inversion relates to uniqueness. Because this is an optimization problem, the solution may be trapped in a local minimum. One may attempt to avoid this pitfall by adopting the so-called multi-scale approach (e.g., Bunks et al 1995; Sirgue and Pratt 2004; Ravaut et al 2004; Fichtner et al 2009) which involves taking the following precautionary steps:

- Choosing a "good" initial model is crucial since meaningless local optima may attract and trap the solution. In the case of sunspots, one may construct 3-D models that are constrained by the surface field.
- Employ travel times of long-wavelength waves (i.e., high phase speeds) that are primarily sensitive to coarse-grained features of the object in question and iteratively refine the model by gradually incorporating travel times of smaller wavelength waves (lower phase speeds).
- Use different types of measurements, i.e., a variety of time-distance averaging geometries, frequency, and phase-speed filters, in the misfit function.
- Ensure a good match between simulations and observations at the photospheric level (e.g., photospheric sunspot magnetic fields or Doppler measurements of surface supergranulation).

Chapter 4
Full Waveform Inversion**

Inferring interior properties of the Sun from photospheric measurements of the seismic wavefield constitutes the helioseismic inverse problem. Deviations in seismic measurements (such as wave travel times) from their fiducial values estimated for a given model of the solar interior imply that the model is inaccurate. In this section, we implement non-linear inversions, executed iteratively, as a means of inverting for the sub-surface structure of perturbations. The model can be successively improved using either steepest descent or Krylov-subspace techniques such as conjugate gradient, or limited-memory Broyden-Fletcher-Goldfarb-Shanno method (L-BFGS; Appendix A.6). For the sake of simplicity in illustrating the method, we consider two distinct 2-D inverse problems one where we attempt to recover a sound-speed perturbation and another where we image flows.

Full waveform inversion (FWI) is a label for techniques widely used in terrestrial and exploration seismology to infer the structure of the highly heterogeneous Earth. The name derives from the goal of fitting the entire waveform by the end of the inversion so that all available seismic information is utilized. It does not necessarily mean that the entire raw waveform is used during the inversion. Specifically, it is found that parametrizing the waveform in terms of classical or instantaneous travel times or amplitudes is an effective strategy towards fitting the entire waveform (as opposed to using the raw waveform itself, e.g., Bozdağ et al 2011; Zhu et al 2013; Hanasoge 2014b). In this chapter we restrict ourselves to classical travel times of well defined parts of the waveform (such as the first or second bounce and filtered times). Thus the method we are discussing is a subset of a larger collection of techniques termed FWI and hence we refer to it as such.

A waveform can be broken up into frequency bands. The full waveform approach involves assimilating all of these measurements into the inversion in the maximally leverage seismic data. A number of inversion methods already adopt aspects of this

---

** The material for this section is primarily taken from Hanasoge and Tromp (2014) and Hanasoge (2014a).

approach (e.g., Švanda et al 2011; Jackiewicz et al 2012; Dombroski et al 2013), strictly assuming however that seismic measurements depend linearly on interior properties. In the present formulation, we compare waveforms solely in the sense of travel times. Further, because we only consider sound-speed perturbations and flows here, the primary impact on waveforms is to shift their phases and to a lesser degree, amplitude. In principle, we may also include amplitudes, instantaneous phase, or even raw waveform differences (e.g., Dahlen and Baig 2002; Bozdağ et al 2011; Rickers et al 2013).

The basic goal in seismology, as discussed in earlier sections, is to relate properties of the interior to wavefield measurements at the bounding surface. The first step involves defining a misfit or cost functional that comprises some measure of the difference between measurement and prediction. An example of a misfit function (χ) in the case of time-distance helioseismology is the L_2 norm of the difference between measurement (τ^o) and prediction (τ) at some set of locations i (Section 3)

$$\chi = \frac{1}{2}\sum_i (\tau_i - \tau_i^o)^2. \tag{4.1}$$

A more general formulation to include a noise-covariance matrix in the definition of the misfit is discussed in Section 3, specifically in Equation (3.4). Here, we study a simpler problem where the data are known exactly, i.e., the noise level is zero. The next step is to determine how to change the model so that the predicted travel times τ_i are closer to the measurements τ^o in the sense of norm (4.1). This is a high-dimensional inverse problem, since we seek to alter various properties such as flows, sound speed, and density of the 3-D interior, thereby introducing a large number of parameters, in order to appropriately alter the travel times measured at the bounding surface of the Sun.

The misfit function (4.1) depends on the model, i.e., $\chi = \chi(\mathbf{m})$, where $\mathbf{m} = \mathbf{m}(\mathbf{x})$ is the model of the Sun and $\mathbf{x}$ is the spatial coordinate. To vary the misfit, we consider the Taylor expansion of equation (4.1) around model $\mathbf{m}$,

$$\delta\chi = \sum_i (\tau_i - \tau_i^o)\frac{\partial \tau_i}{\partial \mathbf{m}}\delta\mathbf{m}, \tag{4.2}$$

and it is seen that to reduce the misfit, i.e., to induce $\delta\chi < 0$, we first need access to the gradient of the misfit function $\partial \tau_i/\partial \mathbf{m}$. Gradient-based optimization methods are designed to address this question, specifically to minimize penalty (4.1), an inherently non-linear function of the 3-D model parameters. The gradient of misfit (4.1) with respect to model parameters is the so-called sensitivity kernel, alternately known as the Fréchet derivative,

$$\frac{\partial \tau_i}{\partial \mathbf{m}} = \mathbf{K}(\mathbf{x}, \mathbf{x}_i; \mathbf{m}), \tag{4.3}$$

where $\mathbf{K}$ is the sensitivity of travel time τ_i to changes in the model $\mathbf{m} = \mathbf{m(x)}$, and is therefore a function of the model and space. Equation (4.3) along with (4.2) gives us a prescription to compute a model that minimizes the misfit for the quiet Sun,

$$\delta\chi = \int_{\odot} d\mathbf{x}\, K_c\, \delta\ln c + K_\rho\, \delta\ln\rho + \mathbf{K_v} \cdot \delta\mathbf{v}, \qquad (4.4)$$

where c is sound speed, ρ is density, and $\mathbf{u}$ are flows, K_c, K_ρ, and $\mathbf{K_v}$ are kernels for sound speed, density, and flows, respectively, (Hanasoge et al 2011, 2012a). We use log quantities for variations in c and ρ since they are positive definite.

Seismic inversions are matrix-inverse problems of the form

$$A\,\delta\mathbf{m} = \{\delta\tau\}, \qquad (4.5)$$

where $A = A(\mathbf{m})$ is a fat matrix of dimension $N \times M$, and where the M unknown model parameters are substantially larger than the N measurements, $\delta\mathbf{m}$ is the model update vector, of size $M \times 1$ and $\{\delta\tau\}$ is an $N \times 1$ vector composed of the travel times. The matrix A comprises the sensitivity of the travel time to model parameters, i.e., it is composed of sensitivity kernels. Sensitivity kernels using only 1-D vertical stratification are easy to construct (Birch et al 2004) and lead to lateral (horizontal) translation invariance, which in turn makes the inversion tractable (Švanda et al 2011). Although likely erroneous for certain problems, this approach is generally invoked regardless because a viable methodology to fully account for the three-dimensionality and non-linearity of the inverse problem has only recently been introduced (Hanasoge et al 2011). Inverse approaches that rely on translation invariance possess the additional feature that the computational cost scales very weakly with the number of measurement points, unlike in the adjoint method. On the other hand, it is possible to mitigate the computational cost of adjoint method based approaches by choosing a set of observation points such that coverage and resolution are maximized.

Matrix A can be very big (with 10^{12} elements or more), and will possess a high condition number, and therefore inverting it is not an option. Consequently, we use an iterative procedure to arrive at some appropriate inverse of A and therefore, $\delta\mathbf{m}$. To perform iterations, a local linear approximation is invoked, much as in the style of the Taylor expansion in equation (4.2), and methods such as steepest descent, conjugate gradient, or the quasi-Newton limited-memory BFGS are applied.

Iterative inversions have the benefit that the misfit in a wide variety of categories such as travel times measured in f or p_1 or with other phase- and frequency-filtered data, can be monitored. A serious drawback of prior flow-inversion testing is that the misfit is never studied post inversion, making the current approach very attractive.

Models of the solar interior are functions of space and are high-dimensional quantities. For instance, in the problem considered here, some 120,000 grid points are used to resolve wave propagation and therefore at least as many parameters. One can therefore consider a *distribution* of models, described by some probability density function and a given model being one realization drawn from this distribution. For each model, there exists a corresponding wavefield which in turn implies one value of the misfit. Thus a high-dimensional quantity is mapped on to one number and it is the task of inverse theory to converge on the 'correct' model that fits the observations. In other words, there is a model that possibly corresponds to a global minimum in misfit that we must find. However, there may also be a variety of local minima in this misfit-model space and it is conceivable that a poor initial guess could lead to the system being trapped in a local minimum. This discussion points to the concept of *model uncertainty* implying that in addition to uncertainty in data (owing to stochastic wave excitation noise), there is a set of models consistent with measurements.

Inversion strategy (a schematic of which is shown in Figure 1.10) consists of making a series of choices that limit the likelihood of being trapped in a local minimum. This is especially important in exploration seismology where models of the oil reservoir can exhibit strong local heterogeneities. The Sun, a convecting fluid, is well mixed and consequently, the issue of strong heterogeneities is not a serious issue (with the exception of sunspots) and model uncertainty is generally not perceived to be very important. The acoustic sound speed plays an overwhelmingly important role in wave propagation and therefore, structure inversions have been observed to be robust to model uncertainty (Hanasoge and Tromp 2014). In other words, for structure-related anomalies, the model-misfit space is such that convergence is likely. However, we demonstrate here that flow inversions are not easily tractable. Depending on the strategy adopted, i.e., type of measurements assimilated, preconditioning applied to the kernels, etc., a range of models show agreement with measurements and the misfit is seen to smoothly fall in various categories.

The adjoint method, a means of obtaining gradients of the misfit function χ, is well studied in the regime of relatively strong heterogeneities, as demonstrated by the successful application to terrestrial seismic inversions of, e.g., the Southern-California crust (Tape et al 2009), European structure (Zhu et al 2013), and Australia (Fichtner et al 2009). This technique is applied to constrained optimization problems in which we seek to minimize the misfit with the constraint that the wavefield satisfy the partial differential equation that governs wave propagation in the Sun.

The following summarizes the steps involved in FWI for the test problems studied here (also see Figure 1.10)

- Construct a true model of the perturbation and compute the associated wavefield at the surface (which we shall term 'observations' here),
- Choose a set of optimally placed sources and a broad set of receivers, since the computational expense scales with the number of sources (Hanasoge and Tromp 2014),
- Determine the surface wavefield for a given model of the solar interior using equation (4.7) and compute the predicted wavefield (the forward calculation),
- Choose which measurements to use in the inversion: low-frequency, large-wavelength modes at the start, gradually introducing higher-frequency data,
- Compute the misfit between predicted and observed data,
- Sum over the gradients (kernels) between every source-receiver pair weighted by the associated travel-time misfit,

- Compute the gradient of the misfit with respect to the model parameters using the algorithm described in Section 3,
- Perform a line search to determine the update that results in the greatest misfit reduction,
- Update the model and repeat.

4.1 Sound-speed Perturbation

This section aims to introduce the basic concepts of this inverse methodology and is not exhaustive in its scope. In the first part of this discussion, we limit ourselves to the study of a sound-speed inversion, described thus

$$\delta\chi = \int_{\odot} d\mathbf{x}\, K_c\, \delta \ln c. \tag{4.6}$$

To compute the misfit gradient K_c, we apply the adjoint method described by Hanasoge et al (2011), used to simultaneously construct kernels K_c, K_ρ, and $\mathbf{K_v}$. However, we only retain K_c for this problem.

We define a simplified helioseismic operator,

$$\rho\partial_t^2\boldsymbol{\xi} = \mathbf{V}(\rho c^2 \mathbf{V}\cdot\boldsymbol{\xi} + \rho g\xi_z) + \mathbf{g}\mathbf{V}\cdot(\rho\boldsymbol{\xi}) + \mathbf{S}, \tag{4.7}$$

where density is denoted by $\rho = \rho(\mathbf{x})$, sound speed by $c = c(\mathbf{x})$, gravity by $\mathbf{g} = -g(z)\hat{\mathbf{z}}$, the vector acoustic wave displacement by $\boldsymbol{\xi} = \boldsymbol{\xi}(\mathbf{x},t)$, whose vertical component is ξ_z, the source by $\mathbf{S} = \mathbf{S}(\mathbf{x},t)$, and time by t. The covariant spatial derivative is denoted by $\mathbf{V}$ and the partial derivative with respect to time is ∂_t. The adjoint method relies on making predictions and using the difference with observations to drive changes in the solar model. Here we use SPARC (Section 2.5) to numerically solve equation (4.7).

The adjoint method consists of computing *forward* and *adjoint* wavefields. The *forward* calculation is a predictor step, making a prediction on the photospheric cross correlation (or some other measurement) along with the attendant 3-D seismic wavefield in the interior. This calculation captures the connection between the interior sensitivity of the wavefield and the surface seismic signature. The *adjoint* calculation consists of performing a 3-D wavefield simulation driven by the difference between prediction and observation, as measured by equation (4.1). Roughly speaking, this captures the connection between the interior and the measurement misfit as recorded at the surface. Finally, the time-domain convolution of forward and adjoint wavefields gives the total misfit gradient, i.e., all the desired sensitivity kernels (Eq. [4.4]). Because this formulation of the adjoint method is numerical, forward and adjoint simulations may be carried out for arbitrary backgrounds. Further, with a few calculations, all relevant kernels may be simultaneously obtained. The analysis, kernel expressions, and algorithm are discussed in Section 3. Finally, we note that the extension to a variety of other measurements such as resonant frequencies closely follows the analysis in section 4 of Hanasoge et al (2011), with the relevant measurement framed in a manner so as to connect it to Green's functions of the medium.

Waves in the Sun are excited in a thin near-surface radial envelope (e.g., Stein and Nordlund 2000) but uniformly in the lateral (horizontal) direction. Thus the helioseismic wavefield is excited by distributed sources, which, together with the stochastic nature of the excitation, makes the calculation of sensitivity kernels complicated (Section 3; also see, Hanasoge et al 2011). This is because the wavefield measured at a given point consists of contributions from a wide range of sources and the cross correlation of the wavefield measured at a point pair thus averages these contributions. However, in the case where the distribution of sources is uniform, the cross correlation can be shown to be closely related to Green's function of the medium (e.g., Snieder 2004). This correspondence allows for treating the second-order cross correlation measured between a point pair as arising from a deterministic, single source-receiver configuration, greatly reducing the complexity of the problem (the point-pair map on to the source and receiver). While it may appear that the solar wavefield is an ideal fit for this correspondence (owing to the lateral uniformity of sources), the damping mechanism and the line-of-sight nature of observations diminish the accuracy of the relationship (e.g., Gizon et al 2010). However, it still serves as a very useful first approximation to study the simplified deterministic source-receiver problem since it allows for the appreciation and development of inverse methodology prior to comprehensive modeling. Kernels in this limit treat each branch of the cross correlation measured between a pair of points as the wave displacement due to a deterministic single source.

4.2 The inversion

Here we discuss practical issues and the choices we have made. We do not start from a vacuum, and indeed, there exists significant geophysical seismic literature on these topics, and the choices from these articles guide our thinking. However, the helioseismic inverse problem possesses its own idiosyncrasies and to optimize our methodology, an exhaustive survey of these choices will be necessary. This is especially the case when including more parameters such as flows and magnetic fields.

4.2.1 True and starting models

The goal is to invert for the true anomaly in sound speed shown in Figure 4.1. Also shown in Figure 4.1 is the starting model, which is a solely vertically stratified, convectively stabilized form of model S (Christensen-Dalsgaard et al 1996; Hanasoge 2007; Hanasoge et al 2008). Sound-speed perturbations shown in Figure are measured as deviations from this 'quiet Sun' stratification, i.e., $[c(x,z) - c_q(z)]/c_q(z)$, where c_q is the nominal sound speed in the quiet Sun and $c(x,z)$ is the sound speed of the current model. To accelerate convergence, we may also constrain the surface layers in the starting model to be identical to those of the true model, the argument being that the surface layers of the true model would be 'observable' (which we do in Section 4.2.10). For now, we choose the starting model, $c(x,z) = c_q(z)$. In the subsequent discussion and in various Figures and attendant captions, we will make use of the following definition

$$\delta \ln c = \ln \frac{c(x,z)}{c_q(z)}. \tag{4.8}$$

4.2.2 Sources and Receivers

Tromp et al (2010) and Hanasoge et al (2011) showed that the cost of inversion scales with the number of sources and hence the nomenclature. Thus having selected points at which to place sources (source pixels), we may increase the number of receivers arbitrarily without accruing additional computational cost. Choices for source pixels are therefore crucial since we would like to maximize seismic information. There are likely more formal and rigorous ways to make this choice but in the effort here, we have discovered through the process of trial and error that placing sources in the near field of the perturbation leads to faster convergence. We thus choose 7 sources placed at points along the sound-speed perturbation as shown in Figure 4.1. In order to introduce more seismic information, we perform a few iterations for a given set of sources and replace these by another set. In the inversion presented here, the sources change from the originally chosen set (indicated by triangles in Figure 4.1) to another set of 7 pixels at iteration 7, indicated by asterisks. The new set of pixels is more sparsely distributed and is spread out over a larger horizontal distance, to improve the imaging aperture. We do not introduce further changes to the set of sources because seismic information is concentrated in the vicinity of the perturbation, which we explore thoroughly with the overall set of pixels. Receivers may also be changed from iteration to iteration, but here, we have maintained the same set of receivers throughout the inversion.

4.2.3 Measurements

We measure wave travel times between point pairs. Using the definition of the linear travel time as set out by Gizon and Birch (2002), we formulate the adjoint method for this measurement (Hanasoge et al 2011). In practice, the relative travel time between two waveforms is measured by actually cross correlating them and extracting the time lag associated with the peak correlation coefficient. For instance,

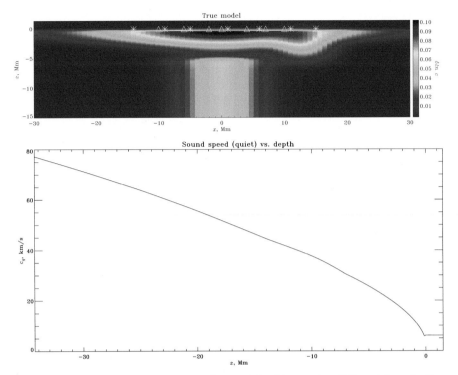

Fig. 4.1 True model (upper panel), where $\delta \ln c$ is defined in equation (4.8), and the quiet-Sun sound speed, $c_q(z)$ in the lower panel. The triangles denote the first set of sources and the asterisks the second set. The sources are switched at iteration 7, to introduce new seismic information. Because wave excitation occurs in the very near-surface layers of the Sun ($z = -50$ km), we fix the location in depth but are free to vary the horizontal location.

if waves appear at point B at a positive time lag in relation to point A, then point B acts as the receiver to source A. In Figure 4.2, we show the time-distance diagram for a source at $x = -15$ Mm. We measure travel times for p-modes over a range of point-pair distances for the first, second, and third bounces over specified frequency bands.

4.2.4 Adjoint source

For a given source point, we measure travel times at receivers located farther than 15 Mm from it. This minimum separation allows for the distinction between the various bounces of p-modes. At distances shorter than 15 Mm, it is no longer possible to clearly interpret the measurement. We only simulate for 1.5 hrs of solar time, which places a restriction on a maximum source-receiver distance possible for each bounce. In the adjoint calculation, the wave equation is forced with *adjoint*

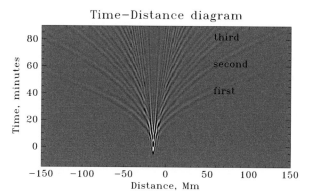

Fig. 4.2 Time-distance diagram. The source pixel in this case is placed at $x = -15$ Mm. Travel-time shifts measured at receivers for a given bounce (first, second, or third) are used in the inversion. In order to distinguish between the various arrivals, we select receivers that are at a minimum distance of 15 Mm away from the source for the first and second bounces and 30 Mm for the third bounce.

sources placed at all the receiver locations where measurements are made. The adjoint source at any given measurement point consists of the travel-time shift multiplied by the time reverse of the temporal derivative of the measured waveform from the forward calculation. In Figure 4.3, the full adjoint source is shown in the upper panel and a cut at a fixed spatial location is shown in the bottom.

4.2.5 Discrete adjoint method

In the formulation adopted here, the adjoint method is treated in a continuous sense (Hanasoge et al 2011), and expressions for kernels that are computed by convolving the forward and adjoint wavefields are derived for continuous space. However, numerical simulations are performed on discrete grids, and indeed, errors are generated when the continuous adjoint formulation is discretized. The gradient thus obtained is not as accurate as when the problem is posed consistently in the discrete sense. This slows down convergence and is a well noted issue in these seismic inverse problems (for airfoil design, see, e.g., Giles and Pierce 2000). Nevertheless, because convergence is observed and because there is no easy or obvious route to a discrete adjoint formulation, we proceed with the (inaccurate) continuous analog.

4.2.6 Preconditioning and Smoothing

While adjoint methods may not explicitly state the role of regularization, it does make its way into the heart of the problem. At every iteration, the total misfit

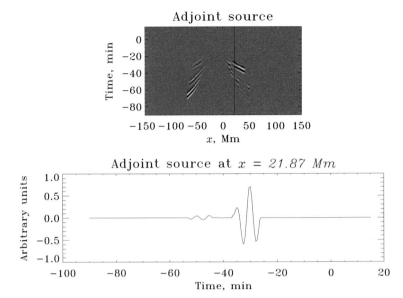

Fig. 4.3 Adjoint sources at receivers (upper panel) corresponding to the source shown in Figure 4.2. Each adjoint source is the time-reversed temporal derivative of the waveform measured at that receiver, multiplied by the cross correlation travel-time shift. The adjoint source at a specific x location is shown in the lower panel. The waveform multiplied by the travel-time shift is the largest for the first bounce, which, owing to time reversal, appears at a later time in the adjoint source. The adjoint source suggests that the most significant travel-time deviations are recorded by the first bounce, thereby playing a prominent role when constructing the gradient.

gradient, summed over all sources, contains non-smooth variations co-spatial with source locations, which may slow convergence. To mitigate this problem, spatial smoothing must be applied to the gradient.

The rate of convergence can be improved by 'preconditioning' the gradient, which in practice involves multiplying the gradient by a suitable function termed the preconditioner, i.e., the gradient is preconditioned first and spatially smoothed next. The sensitivity of the convergence rate to different types of preconditioners was studied by Luo et al (2013), who found that the optimal preconditioner for the problem they were studying was a convolution of the time derivatives of the forward and adjoint wavefields (see their Eqs. [108] and [109]). However, we found that preconditioning (based on the methods of Luo et al 2013) and smoothing led to slower convergence rate in comparison to just smoothing. For the problem of thermal imaging, we restrict ourselves only to smoothing the gradient here. Note that explicit regularization terms (user prescribed) may indeed be included in the original statement of the problem, since the adjoint method is designed to address constrained-optimization problems (Figure 4.4).

Unsmoothed kernel, iteration 1

Smoothed kernel, iteration 1

Fig. 4.4 The raw sound-speed gradient, shown in the upper panel has sharp variations due to numerical issues related to the spatially localized forward source. The smoothed kernel is shown in the lower panel, where a 3-point Gaussian filter was applied to accomplish smoothing. The update is then computed through $c_{02} = c_{01}(1 + \varepsilon \bar{K}_{c_{01}})$, where the overbar indicates smoothing, c_{02} is the sound-speed model for the second iteration and ε is a small constant.

4.2.7 Model updates

Given the gradient, the model can be updated using a variety of methods. The first iteration relies on steepest descent, in which the update is tangent to the gradient direction. At higher iterations, we may choose between conjugate gradient and L-BFGS to create updates. Conjugate gradient requires the previous and current gradients to form the update where L-BFGS can be designed to use the full history of gradients and models to create an update. Although not shown here, from preliminary testing we find that L-BFGS and conjugate gradient converge at roughly the same rate. More careful testing may reveal the parameter regimes where one method is faster than the other.

Since we only consider sound-speed perturbations, the smoothed sound-speed sensitivity kernel is first normalized by its largest absolute value so that it ($\bar{K}_{c_i}$) spans the range $[-1, 1]$. We then perform a line search, using 5 different models, $c_{i+1} = c_i(1 + \varepsilon \bar{K}_{c_i})$, where c_i is the model at the ith iteration, ε is a small constant that takes on values $[0.01, 0.02, 0.03, 0.04, 0.05]$. Every value of ε leads to a model c_{i+1}, and we estimate the misfit for each. At every iteration, we test for local convexity by performing a line search. Typically an elegant parabolic curve is observed, as in Figure 4.5. We choose the model corresponding to the minimum point of this curve as the model for the next iteration, i.e. the update corresponds to the valley of the line search curve. The update parameter ε generally decreases with iteration, and ε for updates to successive models is smaller in magnitude. Typically, $\varepsilon \sim 0.06$ for the very first iteration and then drops to about $\varepsilon \sim 0.004$ at the eleventh iteration.

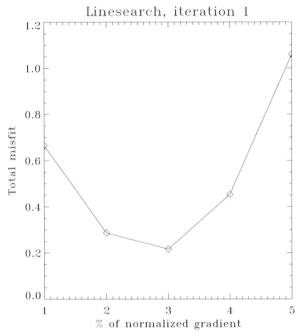

Fig. 4.5 Line search at each iteration to determine ε for the update $c_{i+1} = c_i(1 + \varepsilon \bar{K}_{c_i})$. The x axis shows different values of ε and the y axis the misfit associated which the corresponding model. In this case, we choose the model for which the misfit reaches a minimum, i.e., for $\varepsilon = 0.03$.

Every few iterations, the parabolic line search curve for a non-steepest-descent method is not easily produced. In such scenarios, we revert to steepest descent as a means of 'resetting' the inversion. For instance, we might have the following configuration of updates - 1 - steepest, 2, 3, 4 - conj. grad., 5 - steepest, 6, 7 - conj. grad, where the numbers indicate the iteration index. We show 12 iterations of an inversion for the setup discussed in Figure 4.1 using a combination of conjugate gradient and steepest descent methods in Figure 4.6. We also applied the L-BFGS algorithm after 4 iterations of steepest descent but found the rate of convergence to be generally unchanged. The performance of the method appears to be less sensitive to these choices and much more to the introduction of external information (such as surface constraints, new pixels, etc.).

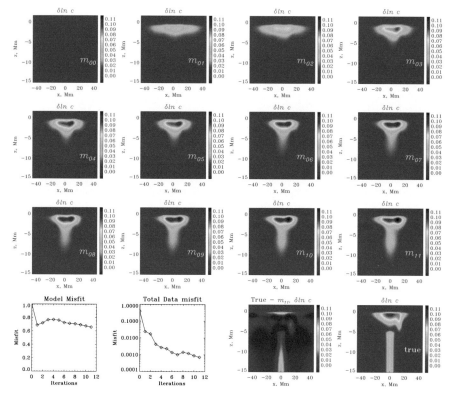

Fig. 4.6 Iterations in a conjugate-gradient based inversion. The first iteration is performed using steepest descent and a combination of conjugate gradient and steepest descent is used to compute subsequent models. At iteration 7, we change the set of sources and this creates a local jump in the data misfit because more information has been introduced. It is seen that models approach the true anomaly gradually but the reduction in both data and model misfits slows down with iteration. The model misfit is the normalized L_2 norm difference between the true and current model whereas the total data misfit is the same as equation (4.1). In the first few iterations, the model misfit increases because surface layers contain significant errors and p-modes possess limited sensitivity to these layers. As the model evolves it overcomes this local hill, appearing to 'fix' the surface layers, and a steady decline is seen in the last few iterations.

4.2.8 Uniqueness

In high-dimensional inverse problems, the choice of the starting model and type of measurements introduced to update the model may be critical to avoiding being trapped in a local minimum. A standard strategy applied to mitigate the chances of encountering this undesirable outcome is to first use measurements taken from low-frequency modes and gradually introduce higher frequencies as the model iteratively accrues features. This particular issue can be very serious when attempting to image reflectors in the interior, as in exploration geophysics, but it is unlikely to be critical for helioseismology. Because the frequency range of trapped modes in the Sun is so narrow (2.5–5.5 mHz), we choose here to utilize the entire passband. Indeed, we are aware that this strategy may not be optimum for all applications but we find it to be successful in the case of sound-speed perturbations studied here.

4.2.9 Testing convergence

To verify that misfit is being minimized for all the measurements, we measure the misfit associated with each model for travel times binned into categories by their bounce number (first, second, or third) and frequency band (2.5–4, 2.5–5, 2.5–5.5). Note that we could also have measured the misfit using ridge- and phase-filtering to isolate modes in various parts of the power spectrum but our categories are simpler in this case. Thus we confirm that the misfit is uniformly reduced in these 9 categories. A similar strategy has been used successfully in terrestrial applications, e.g., Zhu et al (2013) although because terrestrial seismic waves exhibit a larger temporal frequency range, geophysicists apply frequency filters to their data. Fixing the lower frequency cutoff, Zhu et al (2013) increase the upper corner of the bandpass with iteration, gradually allowing in more information as the model grows in complexity. We also calculate the model misfit by computing the L_2 norm of the difference between the true and inverted models as a function of iteration. Both data and model misfit are seen to decrease with iteration in Figure 4.7.

4.2.10 Including "surface" constraints

The sound-speed anomaly studied here has a 'surface' signature and we can include this as a constraint on the model. It is of relevance because in reality, perturbations such as supergranules, meridional circulation, sunspots, and active regions are optically observed at the photosphere and these observations can be used to accelerate convergence. For the inverse problem at hand, p-modes are used to image the sound-speed perturbation. Surface-gravity f-modes, which are very sensitive to the surface, do not register sound-speed perturbations since the restoring force for these waves is gravity and not pressure. Consequently, adding a surface constraint to the inversion is likely to accelerate convergence for this inverse problem.

In Figure 4.8, we see direct evidence of this, where the bottom-left panel shows a smooth decline in model misfit with iteration, unlike in Figure 4.6, which displays a non-monotonic trajectory. Overall, both data and model misfit are lower in Figure 4.8 in comparison to Figure 4.6. We also overplot all the misfit categories in Figure 4.9 to highlight the (anticipated) superiority of surface-constrained inversions.

Finally, we show the improvement between waveforms derived from "data" and the model in Figure 4.10. By iteration 11, the waveforms start matching up well.

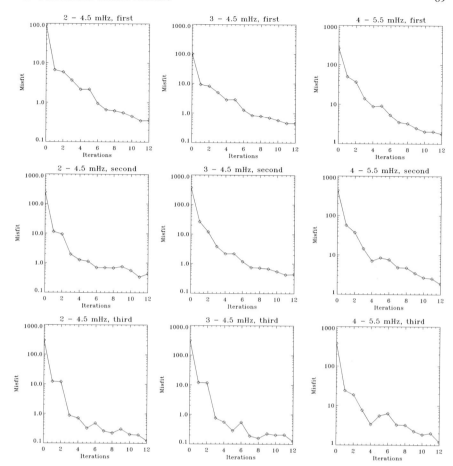

Fig. 4.7 Misfit reduction with iteration, broken up based on the frequency bands and bounces. It is seen that regardless of the band, the misfit decreases uniformly (straying from monotonic reduction along the way on a few occasion). Note that we do not apply a frequency filter in our travel-time measurements, so we are not explicitly attempting to minimize these separate bands. This trend occurs organically, suggesting that the eventual result will be consistent with the governing wave equation and the measurement technique. It also adds support to the notion that the adjoint method in conjunction with linear algebraic inverse methods can be very successful. Note that we could also have used ridge- and phase-speed filtering to further test for a decreasing misfit with iteration.

4.3 Conclusions: thermal structure

Full waveform inversion provides a means of addressing longstanding problems in helioseismology. It directly addresses the major issue of non-linear dependencies of travel times on properties of the solar medium in structures such as sunspots and supergranules. While iterative inversions are indeed possible using ray theory as the forward model, wave propagation is demonstrably not well captured in this

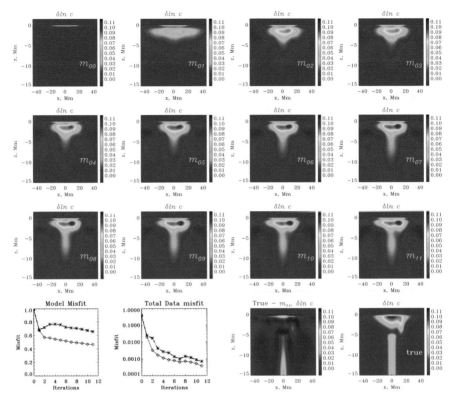

Fig. 4.8 Iterations in a conjugate-gradient based inversion. The starting model contains a 'surface constraint,' as seen in m_{00}. The rest of the algorithm is unchanged from the example shown in Figure 4.6. The first iteration is performed using steepest descent and a combination of conjugate gradient and steepest descent is used to compute subsequent models. It is seen that models approach the true but the reduction in the misfit slows down with iteration. The model misfit is the normalized L_2 norm difference between the true and current model whereas the total data misfit is the same as equation (4.1). For comparison, we overplot the misfit evolution for the unconstrained inversion (dot-dashed line with asterisks). For categories of model and data misfit, it is seen that surface constraints accelerate convergence.

high-frequency approximation (Birch et al 2001). Helioseismology is increasingly a high-precision science and to make accurate inferences, it is important to model wave effects as fully as possible. Finite frequency forward calculations of the helioseismic wavefield are now routinely performed, and we discuss full waveform inversion strategies within this context.

A basic lacuna of current approaches to 3-D helioseismic inversions is that there is rarely a consistency check of how much the inverted model reduces the misfit between seismic prediction and observation. At each iteration in our inversion, we perform a line search to determine how much to change model, and generally find that beyond 3–5% the misfit actually rises, suggesting that the linear connection between misfit and model change is restricted to this regime. Of course, the caveat in

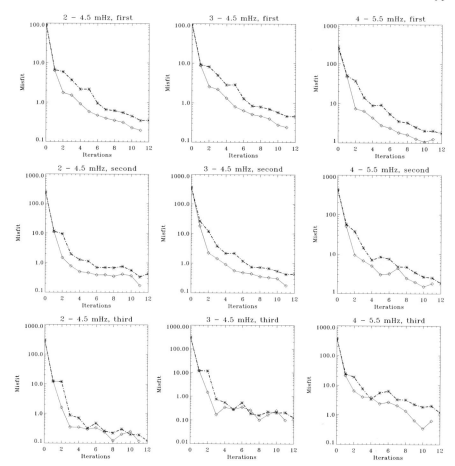

Fig. 4.9 Comparison of misfit bands between surface-constrained and unconstrained inversions. Systematically, unconstrained inversions show slower convergence, as evidenced by the curves with higher misfit (dot-dashed lines with asterisk symbols). Smooth lines with circle symbols show the misfit evolving with iteration for surface-constrained inversions.

drawing this conclusion is that our inversion method is either quasi-Newton- or conjugate gradient based, whereas prior helioseismic inversions have relied on Gauss-Newton-based approaches. In general, Gauss-Newton allows for taking larger steps in model space but it must be emphasized again that the actual extent to which misfit is reduced has generally not been measured. The closest to a consistent inversion can be attributed to Cameron et al (2008), who attempted to study a set of sunspot models using linear magneto-hydrodynamic numerical simulations to determine how well observations can be matched. In a purely forward approach ("probabilistic"), the model space is exhaustively searched, determining the misfit for each model. However, given the computational expense for full wave modeling codes, this may be an infeasible approach.

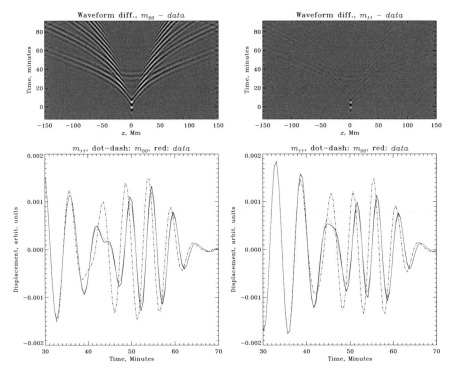

Fig. 4.10 Waveform matching as a function of iteration; difference between time-distance diagrams of models m_{00}, m_{11}, and target data (upper panels). At iteration 11, the difference is substantially smaller (plotted on the same scale). Lower panels show waveforms at $x = -9$ Mm (left) and $x = 22$ Mm (right). By iteration 11, the waveforms match the data very well.

The methodology discussed here still requires development and a more careful exploration of techniques that can enhance convergence. Purely computational test problems, such as the inversion for flows and magnetic fields, will be the focus of future studies. However, full waveform inversion provides a firm theoretical foothold for a field that has long sought a means to accurately interpret helioseismic measurements. The hope is that, with the simultaneous development of inverse theory and high-fidelity numerical methods to rapidly simulate wave propagation in a medium that closely mimics the Sun, we may finally able to settle issues of great relevance to understanding solar dynamics.

4.4 FWI applied to flows in the interior of the Sun

Models of material flows in the interior of the Sun can assist significantly in our understanding of its dynamics. Consequently, substantial effort has been directed towards seismically imaging flows underneath sunspots (e.g., Duvall et al 1996;

Zhao et al 2001; Gizon 2003; Gizon et al 2009; Jain et al 2012), supergranulation (Duvall and Gizon 2000; Beck and Duvall 2001; Gizon et al 2003; Zhao and
Kosovichev 2003; Birch et al 2007; Woodard 2007; Hirzberger et al 2008; Švanda
2012; Duvall and Hanasoge 2013; Duvall et al 2014), meridional circulation (e.g.,
Giles et al 1997; Gizon 2004; Braun and Birch 2008; Komm et al 2013; Zhao et al
2013), and convection (e.g., Swisdak and Zweibel 1999; Duvall 2003; Hanasoge
et al 2012b; Woodard 2014). Seismic inferences are solutions to inverse problems
of the form $Ax = b$, where b is a set of measurements, A is a matrix comprising of
transfer functions between medium properties x and the measurements. Seismology
is primarily a methodology to measure wavespeeds of the medium through which
waves propagate. Three types of wavespeeds govern helioseismic wave propagation
(Hanasoge et al 2012a): an isotropic sound speed (locally independent of direction),
a symmetry-breaking, anisotropic flow velocity (locally dependent on the angle of
propagation), and an anistropic (symmetry-conserving) magnetic Alfvén velocity.
Because flows can break the symmetry of wave propagation, i.e. waves propagating
along and opposite the flow direction are phase shifted in opposite senses, the typical
measurement is of direction-dependent phase-degeneracy lifting, such as $\tau_+ - \tau_-$,
where τ is the travel time and $\pm$ denote pro- and retrograde directions with respect
to the flow.

The analysis and extraction of seismic data from raw observations, taken for
instance by HMI (Schou et al 2012), is well understood and the techniques are established. However, the interpretation of these measurements to create models of
the solar interior has greatly lagged observation. In the context of flow inversions, a
number of authors have constructed algorithms (e.g., Kosovichev et al 2000; Birch
and Gizon 2007; Jackiewicz et al 2007; Švanda et al 2011; Hanasoge et al 2011;
Jackiewicz et al 2012) and performed synthetic tests (e.g., Giles 2000; Zhao and
Kosovichev 2003; Hanasoge et al 2010a; Dombroski et al 2013; DeGrave et al 2014)
to verify and validate these techniques. These validation tests involved calculating
the seismic response to an input (user-prescribed) flow system and subsequently
inverting the responses to test if the original flow system was indeed retrieved. Unfortunately, the overwhelming fraction these efforts drew the conclusion that seismology was unable to accurately infer the input flow system. However, a number of
tests were inconsistent since the response to the input flow system was only calculated approximately, using either a ray approximation or the same sensitivity kernels
(transfer functions) that are used in the inversion. Further, because all prior local inversions have only consisted of one step, non-linearities between model parameters
and measurements are not accounted for. Finally, few inversions in the past have
considered satisfying mass conservation of the inverted flows.

A highly successful result in helioseismology is the inference of global rotation shear (Schou et al 1998), which has withstood repeated testing and remained
consistent. More complicated flow systems involving lateral and radial flows such
as meridional circulation, convection, etc. present substantive challenges in the
guise of 'cross talk' between vertical and horizontal flows on seismic signatures,
making it difficult to distinguish the two. Repeated tests have shown cross talk

to be generally unavoidable (Zhao and Kosovichev 2003; Dombroski et al 2013). We suggest here that this is due to the poor radial localization prevalent in non-axisymmetric inversions.

The seismic inversion is a projection of a feature (such as a supergranule or a sunspot) onto the basis of eigenfunctions of oscillation modes. For global modes (at low-ℓ), there are a large number of radial orders $n \lesssim 20$, whereas at relatively high-ℓ, the regime of interest here, there are far fewer radial orders to choose from (owing to the acoustic cutoff frequency which sets the maximum allowed temporal frequency of trapped modes to 5.5 mHz). Thus the radial resolution is much finer in global seismology in comparison, inherently placing global inversions on a firmer footing.

A commonly used strategy in terrestrial seismology is to separate measurements by frequency and wavelength. Very early on in the inversion, only low-frequency, large-wavelength modes are used, allowing only coarse changes to model. Once the misfit is sufficiently reduced, then the frequency is increased and relatively small-wavelength modes are introduced, refining the prior model. This process must be controlled carefully since allowing in small-wavelength modes at the very start can lead the result down an incorrect path of (misfit) descent. We test the utility of this strategy as well.

4.5 Formulation

We start by defining the wave equation that will be used to study the toy problem in this section. The equation is similar in form to the operator (4.7), but with the addition of an advection term,

$$\rho \partial_t^2 \boldsymbol{\xi} = 2\rho \mathbf{v} \cdot \nabla \partial_t \boldsymbol{\xi} + \nabla (\rho c^2 \nabla \cdot \boldsymbol{\xi} + \rho g \xi_z) + \mathbf{g} \nabla \cdot (\rho \boldsymbol{\xi}) + \mathbf{S}, \qquad (4.9)$$

where $\mathbf{v} = \mathbf{v}(\mathbf{x})$ is background flow velocity.

In equation (4.9), hydrostatic balance has been already accounted for, which is why background pressure does not make an explicit appearance. Flows are considered to be small perturbations around this hydrostatic state and are assumed to not contribute to the force balance. The only parameters that can be varied in equation (4.9) are density, flow velocity and sound speed. Note also that we do not retain the second-order advection term, $\rho \mathbf{v} \cdot \nabla (\mathbf{v} \cdot \nabla \boldsymbol{\xi})$. As in equation (4.4), variations in the misfit are written therefore in terms of model parameters as

$$\delta \chi = - \int_\odot d\mathbf{x} K_c \, \delta \ln c + K_\rho \, \delta \ln \rho + \mathbf{K_v} \cdot \delta \mathbf{v}. \qquad (4.10)$$

Because c, ρ are positive-definite quantities, we can study normalized variations such as $\delta c / c$ and $\delta \rho / \rho$ (and hence the logarithms) and the kernels are directly comparable. However, because equation (4.10) contains dimensional flow variations

$\delta \mathbf{v}$, it is therefore not directly comparable to the other terms. In a constrained-optimization problem where a variety of terms are competing to explain the misfit, it is important to pose the problem in such a way that all the terms are dimensionally compatible. In order to do so, let us consider the physics of flow advection and how it phase shifts waves.

Advection by a flow $\mathbf{v}$ induces frequency shifts to a wave with wavevector $\mathbf{k}$ thus

$$\delta \omega = \mathbf{v} \cdot \mathbf{k}, \tag{4.11}$$

where ω is the frequency, δ represents a shift in the respected quantity, and $\mathbf{k}$ is the wavevector. Defining τ as the wave travel time, the following approximate relation holds

$$\frac{\delta \tau}{\tau} = \frac{\delta \omega}{\omega} = \frac{\mathbf{v} \cdot \mathbf{k}}{c |\mathbf{k}|} = -\frac{\mathbf{v} \cdot \hat{\mathbf{k}}}{c}, \tag{4.12}$$

where c is the sound speed and $\hat{\mathbf{k}}$ is the normalized wavevector. Thus the Doppler-shift term in the wave operator is

$$\delta \mathbf{L} = -2i\omega\rho\,\delta \mathbf{v} \cdot \nabla, \tag{4.13}$$
$$\nabla \cdot (\rho \mathbf{v}) = 0, \tag{4.14}$$

where constraint (4.14), which enforces mass conservation, must be satisfied. The gradient of the misfit functional in equation (4.3) as computed based on the algorithm described in Hanasoge et al (2011) is composed of the temporal convolution between Green's function and its adjoint. Green's function is the response of the wave operator to a delta source. We denote it by $\mathbf{G} = G_{ij}(\mathbf{x}, \mathbf{x}', \omega)$, where G_{ij} is Green's tensor, i is the direction along which the wavefield velocity is measured, j is the direction along which the source is injected, $\mathbf{x}$ is the receiver, and $\mathbf{x}'$ is the source. Seismic reciprocity is a statement about the quantity $\mathbf{G}^\dagger = G_{ji}(\mathbf{x}', \mathbf{x}, \omega)$ in relation to the original Green's function. For a system with no flows, the following statement is true $\mathbf{G}^\dagger = G_{ij}(\mathbf{x}, \mathbf{x}', \omega)$ (depending on the boundary conditions; Hanasoge et al 2011). However, when there are flows, the statement is $\mathbf{G}^\dagger|_{\mathbf{v} \to -\mathbf{v}} = G_{ij}(\mathbf{x}, \mathbf{x}', \omega)$, which means that the reciprocal Green's function for a system where the flows are reversed in sign is identical to the original Green's function (with the correct sign of flows). It turns out that maintaining this relationship is critical to self-consistent interpretations of travel times in the Sun. This relationship is however only valid when mass conservation is maintained. We therefore seek a formulation where mass is explicitly conserved.

Since we are considering flow inversions in the $x - z$ plane, we may introduce the scalar stream function ψ,

$$\mathbf{v} = \frac{1}{\rho} \nabla \times [\rho c (\psi - \psi_0) \mathbf{e}_y], \tag{4.15}$$

where ψ_0 is some constant fiducial value whose role is to ensure that $\psi(\mathbf{x})$ is a positive-definite quantity. When $\psi = \psi_0$, the flow is identically zero. Recalling that the misfit arising from flow perturbations is

$$\delta\chi_{\text{flow}} = -\int_\odot d\mathbf{x}\,\mathbf{K_v}\cdot\delta\mathbf{v} = -\int_\odot d\mathbf{x}\,\mathbf{K_v}\cdot\frac{1}{\rho}\nabla\times\{\delta[\rho c(\psi-\psi_0)]\mathbf{e}_y\} - \mathbf{K_v}\cdot\mathbf{v}\,\delta\ln\rho.$$

Defining

$$K_\psi = \rho c\psi\mathbf{e}_y\cdot\nabla\times\frac{\mathbf{K_v}}{\rho}, \qquad (4.16)$$

using the vector identity $\mathbf{a}\cdot\nabla\times\mathbf{b} = \nabla\cdot(\mathbf{a}\times\mathbf{b}) + \mathbf{b}\cdot\nabla\times\mathbf{a}$, and noting that we employ zero-Dirichlet boundary conditions (also see, Hanasoge et al 2011),

$$\delta\chi = -\int_\odot d\mathbf{x}\,\delta[\rho c(\psi-\psi_0)]\,\mathbf{e}_y\cdot\nabla\times\frac{\mathbf{K_v}}{\rho} - \mathbf{K_v}\cdot\mathbf{v}\,\delta\ln\rho$$

$$= -\int_\odot d\mathbf{x}\left[\left(1-\frac{\psi_0}{\psi}\right)\delta\ln\rho + \left(1-\frac{\psi_0}{\psi}\right)\delta\ln c + \delta\ln\psi\right]K_\psi - \mathbf{K_v}\cdot\mathbf{v}\,\delta\ln\rho.$$

The first two terms encode the cross talk between density and flow and between sound speed and flow. Redefining kernels for sound speed and density thus

$$K_c \to K_c + \left(1-\frac{\psi_0}{\psi}\right)K_\psi, \quad K_\rho \to K_\rho + \left(1-\frac{\psi_0}{\psi}\right)K_\psi - \mathbf{K_v}\cdot\mathbf{v}, \qquad (4.17)$$

we arrive at a formulation for the flow inversion that simultaneously satisfies mass conversation and is written in terms of non-dimensional variations

$$\delta\chi = -\int_\odot d\mathbf{x}\,K_c\,\delta\ln c + K_\rho\,\delta\ln\rho + K_\psi\,\delta\ln\psi. \qquad (4.18)$$

4.6 Problem Setup

We define the 'true' flow model of supergranulation based on the formula described by Duvall and Hanasoge (2013) and Duvall et al (2014). We consider no sound-speed or other perturbations. The wavefield simulated using this model as measured at the surface of the computational box is termed 'data,' which we use to perform the inversion. The wavefield associated with the sequence of flow models in the iterative inversion is termed 'synthetics.' The goal is to fit synthetics to data by appropriately tuning the flow model. Because we consider neither density nor sound-speed anomalies in the true model, we invert only for flow perturbations. The inverse problem we are solving is

$$\delta\chi = -\int_\odot d\mathbf{x}\,K_\psi\,\delta\ln\psi. \qquad (4.19)$$

The flow model is 2-D and with no loss of generality, we consider a 2-D inverse problem, along the lines of Hanasoge and Tromp (2014). These reduced problems place substantially lighter computational demands and provide insight into inversion strategy.

In order to numerically solve equation (4.7), we employ SPARC (Section 2.5). The cost functional used in these calculations is the L_2 norm difference between predicted and observed wave travel times τ at a number of spatial locations i on the surface,

$$\chi = \frac{1}{2}\sum_i (\tau_i - \tau_i^o)^2, \tag{4.20}$$

where τ_i is predicted and τ_i^o is observed. The goal is to minimize χ knowing that $\tau_i = \tau_i(\xi)$ and because the misfit is dependent on background model parameters $\mathbf{m}$, $\tau_i(\xi) = \tau_i(\mathbf{m})$. Thus the idea is to carefully follow the nested dependencies of wavefield measurements to eventually make the connection to model parameters (i.e., flows in this case).

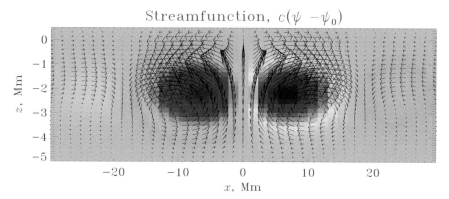

Fig. 4.11 True flow model. The contours show the stream function and arrows indicate the true velocity profile (see Eq. [4.15]). The longest arrow represents a flow speed of 600 m/s. The arrows along the center line are difficult to discern but the maximum vertical flow occurs at $x = 0$, of order 250 m/s. The prescription for this flow is taken from the mass-conserving model discussed by Duvall and Hanasoge (2013).

4.6.1 Supergranulation

We use a 2-D Cartesian computational grid with 512×300 points, spanning $800 \times 138\,\mathrm{Mm}^2$ in the horizontal and vertical directions, respectively. The vertical grid is uniform in acoustic travel time, extending from $r = 0.8R_\odot$ to $r = 1.001R_\odot$. We place the supergranule model in Figure 4.11 at the horizontal center of the domain. We choose seven locations across the supergranule where we excite waves.

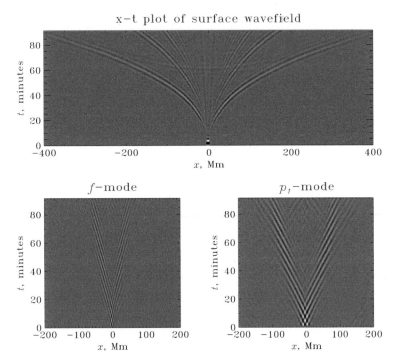

Fig. 4.12 Time-distance plot of the wavefield observed at the surface. The source is placed at $x = -1$ and waves that are generated are observed at various distances and times at the surface. The lower plots shows f- and p_1-mode-filtered surface wavefields. The parabolic feature in the top panel is due to the partial reflection of waves from the bottom computational boundary. This occurs despite the utilization of ostensibly high-fidelity-absorptive perfectly matched layers. Because we use temporal Fourier transforms, the signal wraps around in time as seen in the bottom two panels. This causes some 'noise' in the measurements at late times but is unlikely to play a serious role because the overlap is minor.

Measurements of the wavefield, taken at hundreds of receivers on either side of the source, are used in the inversion. Sources are at a fixed radial location of 100 km below the photosphere and receivers are placed 200 km above, mimicking the excitation and measurement processes in the Sun. We do not fully model cross correlation measurements in this work because of the additional added complexity. Instead, we limit ourselves to deterministic sources, much as in Section 4.1, since our aim is to establish the viability of FWI for flow inversions. In order to start with no bias and conditions similar to the inversion in Section 4.1, we start with model $\psi = \psi_0$, i.e. no flows. Subsequently, we follow the procedure outlined in Figure 1.10. For measurements, we use a continuous span of receivers, starting with f-filtered travel times for source-receiver distances ranging from 8 to 25 Mm, p_1-filtered travel times for distances from 10 to 35 Mm and first-bounce unfiltered p travel times for distances from 35 to 350 Mm (see Figure 4.12). We recognize that it is important to include spherical-geometric effects for such substantial wave travel distances when dealing with observations. However, in this case, the 'data' come from the same numerical code, so the approach is consistent. To reiterate, we use large-distance measurements to test their efficacy in improving the fidelity of inversions. Frequency filters are not applied although some minor experimentation showed benefits to be limited. We precondition the gradient with the approximate Hessian described in Luo et al (2013) and Zhu et al (2013). As advertised, and displayed in Figure 4.13, the misfit decreases uniformly in every measured category, at least for the first several iterations. The evolution of the model of the supergranule is shown in Figure 4.14 and the raw waveform misfit in Figure 4.15. The model is strongly surface peaked, and improvements at depth occur very slowly, and we find that kernels for the inversion have concentrated power in the near-surface layers. A careful examination of the f, p_1, and p kernels (not shown here) reveals that it is difficult to remove the effect of the surface (reminiscent of the 'shower-glass' effect; Schunker et al 2005; although that was applied to magnetic fields). Beyond a certain number of iterations, Figure 4.13 illustrates a tradeoff between f, p_1 and p-modes, signaling incorrect depth localization of the flow model.

It is worth considering why global helioseismic inversions have been successful in inferring rotation, since, ostensibly, global modes have similar systematical biases. A substantial advantage of global modes is their high resolution in radial order and spherical harmonic wavenumber, allowing for the direct manipulation of global-mode kernels in order to diminish the surface tail. In our analysis, we treat travel

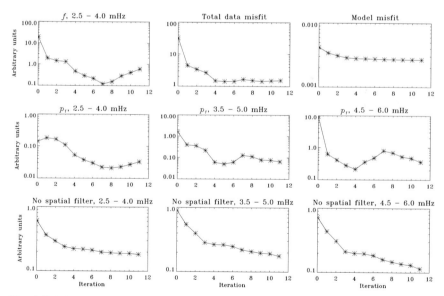

Fig. 4.13 Misfit as a function of category and iteration. The f-mode misfit falls by a factor of almost hundred before rising again, trading off with the p_1- and p-mode misfit. We note that we only use large-distance measurements when estimating the misfit of the lowest panels; for such large distances, the signal is entirely comprised by p-modes. The total data misfit changes very slowly beyond the first few iterations, a manifestation of the tradeoff between different modes. The model misfit, which is defined as the L_2 norm of the difference between the true and inverted models is seen to decrease very slowly.

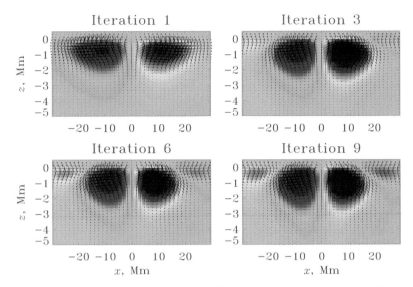

Fig. 4.14 Flow model as it evolves with iteration. The flow model gradually converges to the true model but at a very slow rate. The iterated models peak at the surface whereas the true model (Figure 4.11) peaks at a deeper layer. In the current inversion, errors in vertical and horizontal flow speeds occur due to poor depth localization.

times obtained from the seismic waveform using its entire available bandwidth, i.e. we do not apply frequency filtering (separating measurements in frequency appears to have only minor gains). This results in a diminished frequency resolution in comparison to global modes, contributing to the poor localization in depth.

Nevertheless, these poor convergence properties are in strong contrast with the structure inversions of Section 4.1, where it was demonstrated that with a smaller set of measurements, it was possible to recover the full details of a sound-speed perturbation. One may speculate that flow inversions possess a larger null space but whether this holds water remains to be determined. It is likely that the seismic measurements contain sufficient information to discern between models peaked at different depths, but that the inversion is poorly conditioned, resulting in low convergence rates. We therefore suggest a probabilistic forward search as a means of locating the global minimum.

4.7 Conclusions: flows

Rotation stands unique as a well tested and verified flow system. However rotation is an entirely lateral flow and is axisymmetric. Flows with overturning motions such as meridional circulation, supergranulation, and convection show radial and lateral motions, the latter being non-axisymmetric. Further, meridional circulation is antisymmetric across the equator, making it a difficult target to image using

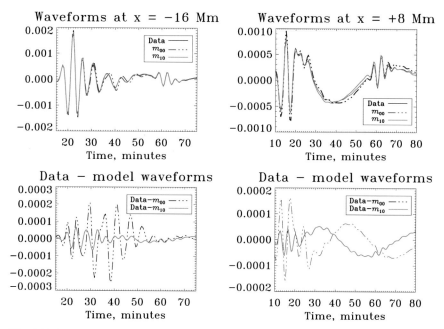

Fig. 4.15 Evolution of the waveform with iteration where the source is at $x = -1$ Mm. Shown are 'data' and waveforms for models at iterations 0 and 10 recorded at different x locations on the surface. Essentially these are cuts at constant x of the time-distance plot (Figure 4.12). Because the data and model 10 waveforms are difficult to distinguish in the upper panels, we show the difference between data and models at iterations 0 and 10 in the bottom two panels for the same x locations. The waveform misfit is seen to reduce but not as significantly as one might hope.

global modes (although unconventional means have been explored by Woodard et al 2013). Consequently, local targeting methods such as time-distance, ring analysis, and holography are necessary. It has however been pointed out that in overturning mass flows, the vertical and horizontal components are not easily distinguished (Zhao and Kosovichev 2003), and therefore it is unclear if even local methods are able to overcome this issue.

We tried two different strategies, one in which all measurements are introduced at the start of inversion and another where only large-distance p-mode travel times are introduced at the first iteration and subsequently, f- and p_1-modes are added. The latter appeared to converge somewhat more rapidly, but neither strategy produced the correct model. The basic error in the model is the inaccuracy in recovering the vertical flow velocity, which was almost a factor of ten smaller than the true model. This occurred because the depth of the inverted flow was not correctly obtained. At the end of the inversion, the overall misfit fell by over a factor of 10 but saturated at this level. Further iterations did not cause the misfit to decrease appreciably, resulting rather in a misfit tradeoff between f and p_1. Thus appears to suggest that the inversion is trapped in some sort of local minimum. The issue may be traced to the fact that for small-scale features, there are relatively few modes that can be used in the seismic analysis (at high-ℓ, the power spectrum shows a sparsity in number of modes). Thus resolving the depth structure of these features is difficult. A corollary to this insight is that it is not evident that even by starting at a model that is very close to the true model, the inversion will push the model towards to the right direction. Duvall and Hanasoge (2013) introduced the idea of using large-distance measurements; such modes ostensibly increase the number of radial orders available to the inversion. However, even this set of measurements was found to be insufficient in the end. In contrast, it must be noted that in Section 4.1, it was possible to recover a thermal structure anomaly using FWI with reasonable accuracy, whereas the flow inversion here has not been nearly as successful. That reason for this dichotomy is not yet fully apparent but one possibility is that flow inversions also have a null-space issue. It is unlikely that increasing the number of observations will improve the convergence rate.

Despite the elaborate nature of this technique, the inversion failed converge to the correct solution. In contrast, conventional, local flow inversions involve only one step, and no independent means of verifying that the flow model explains the seismic measurements better are applied. It is therefore important to estimate the model uncertainty by studying the misfit associated with a class of models. A probabilistic search over a range of forward models, à la Khan et al (2009), for instance, which involves simulating waves through a number of models, measuring the misfit associated with each and locating the deepest minimum, is a useful technique. Additionally, this removes the limitation of trying to project the flow model on the limited set of eigenfunctions available at high wavenumbers and one can use a larger variety of models (also e.g., Cameron et al 2008). Despite the large number of surface observations, it is surprising to note that it is not just realization noise (data uncertainty) that determines the accuracy of imaging flows in the solar interior but also likely model uncertainty.

The implications for the supergranulation models of Švanda (2012), Duvall and Hanasoge (2013), and Duvall et al (2014), which involve large-magnitude vertical and horizontal flows are not entirely clear. While Duvall and Hanasoge (2013) suggest the use of large-distance measurements, which correspond to large-wavelength (coarse-scale) modes, it is not apparent that the eventual solution is accurate.

The problem could also be posed in alternate formulations, such as by connecting the flow to the stream function in equation (4.15) according to

$$\mathbf{v} = \frac{1}{\rho} \nabla \times [\rho c^2 (\psi - \psi_0) \mathbf{e}_y], \tag{4.21}$$

which places greater weight on the inversion in deeper layers ($\propto c^2$). We attempted this approach and while this produces larger vertical velocities, the solution had moved to a different local minimum (and not the global minimum). However this avenue remains to be explored more thoroughly.

Admittedly, this result is discouraging in that even in this idealized inversion, the flow cannot be exactly recovered. Thus, model uncertainty should be considered as a critical part of the inversion. A forward search over a broad class of flow models may be the most productive technique since this would, in addition to potentially discovering the global minimum, allow us to map the model-misfit space. We could then place model uncertainties on the flow inversion, which together with data uncertainty or realization noise, would allow for more accurate uncertainty quantification. An alternate possibility is the use of L_1-norm minimization, in which one might place an additional constraint requiring that the solution to the inverse problem, when projected on a suitable basis, produces coefficients that possess a minimum L_1 norm (Candès et al 2006).

Chapter 5
Whither helioseismology?

High-quality observations of the Sun are widely available, gathered by NASA's space-based flagship heliophysics mission, the Solar Dynamics Observatory (SDO) and by the continuous ground-based monitoring network, GONG. These exquisite observations reveal tantalizingly complex physics at the surface and in the atmosphere, the layers of the Sun that are optically accessible. Whereas, our understanding of processes in the solar interior still remains incomplete. Because the interior cannot be optically imaged, we rely on measuring and interpreting signatures of acoustic waves that propagate within and emerge at the surface. The physics of solar oscillations is well understood and seismic observations can be accurately explained using the theory of linear wave propagation in stratified media. Indeed linear wave physics has been shown to be successful at predicting oscillation frequencies to within parts of a thousand (Christensen-Dalsgaard 2002). This accuracy has allowed for the trustworthy inference of internal differential rotation and structural properties such as the sound speed and composition using measurements of solar oscillation frequencies. That these results have been reproduced by groups around the world (e.g., see, Christensen-Dalsgaard 2002; for a review) using a range of inverse methodologies has strengthened the significance of the inferences. As a consequence, helioseismology has matured into a precision science.

Naturally, seismic methods have turned to the inference of non-axisymmetric features of the solar interior such as multi-scale convection and sunspot structure and dynamics, predicting the emergence of magnetic features, etc., some of which have surface signatures and are therefore observed clearly at the surface. The past successes of global helioseismology were thought to be harbingers of future triumphs with local methods. Unfortunately (or fortunately for it provides a rich source of interesting problems), the promise has not yet been realized, and the central goal of appreciating the dynamics of the non-axisymmetric solar interior remains to be met.

Measuring statistically significant and useful seismic signals from solar data is a problem that was solved decades ago; what has not kept up with the flood of observations and the attendant analyses is a strong theoretical and computational foundation from which to interpret these data. This monograph aims to bridge some

© The Author 2015
S. Hanasoge, *Imaging Convection and Magnetism in the Sun*, SpringerBriefs in Mathematics, DOI 10.1007/978-3-319-27330-3_5

of those gaps, namely a technical introduction to the theoretical and computational methods that are likely to be important to achieving progress. While a suite of inverse methods have been applied in the past to analyze observations and draw inferences (see Gizon et al 2010), the use of more sophisticated techniques such as full waveform inversion remains a frontier research area and one whose potential is not yet realized.

To answer the grand question reflected in this chapter's title, we identify topics that must be investigated in greater detail:

5.1 Forward Model

Waves propagate through layers of convection that grow progressively intense, especially near the surface. In particular, granulation at the surface likely affects mode physics in a systematical way that is not captured in the standard wave equation. Modelling the impact of small-scale stochastic turbulent scattering on waves is a challenging problem (e.g., Bhattacharya et al 2015); deriving an effective equation that accounts for the impact of convection is an important step towards enabling accurate seismic inference. The separation between the wavelengths of typical seismic waves and the scale of granulation allows for asymptotic techniques such as homogenization to be applied.

Keeping this in mind, the design of an accurate forward solver that solves the linear wave equation in a model of the Sun's structure also represents a challenge. In particular, the solar model is convectively unstable and small perturbations will grow exponentially in time without non-linear terms to check the growth. As a consequence, the linear wave equation must be written in temporal frequency domain and solved as a boundary-value problem instead of an initial-boundary-value problem. Boundary-value solvers for large systems of hyperbolic wave equations are still an active area of research; modern methods such as multi-grid show poor convergence properties (Adams 2007). Any hope of obtaining accurate solutions to inverse problems in helioseismology depends on the ability to build robust forward models of wave propagation.

5.2 Inverse theory

Forward model in hand (or once developed), we arrive at a critical step in the process of inference: how to obtain a model of the interior based on seismic measurements at the surface of the Sun. In a typical local helioseismology problem, one can expect to make hundreds of thousands to millions of measurements of wave travel times. These enormous data sizes are a consequence of the resolution of the raw data themselves - the Helioseismic and Magnetic Imager (HMI) beams down 4096 × 4096 sized images every 45 seconds. Even after downsampling and analyzing these time series, the quantity of data is still immense. The data are not all independent - some part of it is redundant; further, the noise associated with different measurements can show significant correlation (Gizon and Birch 2004).

The goal of the inverse problem is to map these data to properties in the interior, such as flows or magnetic fields or thermal structure. Certainly, it is the case that flows such as supergranules and meridional circulation are expressed by a handful of parameters, such as the flow magnitude at the surface (which is known directly from observations), the location at depth corresponding to where the flow magnitude peaks, the location in depth of where the flow turns around (to enforce mass conservation), and parameters that govern the decay of the flow magnitude away from the peak. A similar characterization of other flows and magnetic structures can be applied. Thus we can imagine using thousands of measurements to image ten or fewer parameters - the problem is indeed sparse in this parameter space that describes the flow.

A technique of some interest to attack these problems is basis pursuit denoising, in which the standard least-squares problem is regularized using the L_1 norm of the solution vector. In other words, L_1 regularization enforces sparsity of the solution in the basis of choice (in which it is assumed to be sparse). It has been shown that given certain conditions, the method is able to reconstruct the solution with high probability. Sparse methods have yet to be applied to the helioseismic scenario and they present a promising avenue forward.

Sunspots are strong deviations away from the ambient Sun, and the corresponding seismic shifts are likely to not scale linearly with the magnitude of field strength. More plausibly, there is a non-linear function that relates shifts in seismic measurements to the properties of the sunspot (note that wave propagation is still governed by linear dynamics). Monte Carlo methods that examine the seismic shifts associated with a grid of models are simply too expensive. Each calculation can take thousands of computing processor hours to run. Rather, gradient-based methods in conjunction with the adjoint method to compute gradients provide a useful way to address inversion non-linearity. The design of an optimal strategy that codifies choices for measurements, preconditioning, and smoothing is still an open area and presents a means of gaining insight into the helioseismic inverse problem.

In summary, the hope is that the topics detailed in this monograph present a foundation from which to pursue these lines of future investigation, for, building models of the structure and dynamics of the Sun is a worthwhile proposition that affords the opportunity to appreciate physics in an extraordinary regime.

Appendix A
Interpreting Seismic Measurements

A.1 Seismic Reciprocity

We recall the helioseismic wave operator defined in equation (3.3). The operator may be split into two parts, Hermitian $\mathscr{H}$ and anti-Hermitian $\mathscr{H}^\dagger$, where the former satisfies the following relation

$$\int_\odot d\mathbf{x}\, \boldsymbol{\xi}_A \cdot \mathscr{H} \boldsymbol{\xi}_B = \int_\odot d\mathbf{x}\, \boldsymbol{\xi}_B \cdot \mathscr{H} \boldsymbol{\xi}_A. \tag{A.1}$$

The proof of the self-adjointness of the ideal MHD equations with damping and no background flow is fairly intricate and will not be repeated here. For a thorough demonstration, please refer to, e.g., Goedbloed and Poedts (2004). The only anti-Hermitian part of equation (3.3) contains the background velocity term

$$-2i\omega \int_\odot d\mathbf{x}\, \boldsymbol{\xi}_A \cdot (\rho \mathbf{v} \cdot \boldsymbol{\nabla}) \boldsymbol{\xi}_B = 2i\omega \int_\odot d\mathbf{x}\, \boldsymbol{\xi}_B \cdot (\rho \mathbf{v} \cdot \boldsymbol{\nabla}) \boldsymbol{\xi}_A, \tag{A.2}$$

where the sign of the two integrals upon switching states A and B is reversed. Green's theorem (also Eq. [3.27]) tells us

$$[(\mathscr{H} - 2i\omega \mathbf{v} \cdot \boldsymbol{\nabla})\mathbf{G}(\mathbf{x}, \mathbf{x}_A)]_{ip} = \delta_{ip}\, \delta(\mathbf{x} - \mathbf{x}_A). \tag{A.3}$$

In order to demonstrate reciprocity, we consider another wave state due to a source B, whose Green's function is given by

$$[(\mathscr{H} + 2i\omega \mathbf{v} \cdot \boldsymbol{\nabla})\mathbf{G}^\dagger(\mathbf{x}, \mathbf{x}_B)]_{iq} = \delta_{iq}\, \delta(\mathbf{x} - \mathbf{x}_B). \tag{A.4}$$

Now consider forming the following representation $G_{qi}^\dagger(\mathbf{x}, \mathbf{x}_B) \times (\text{A.3}) - G_{pi}(\mathbf{x}, \mathbf{x}_A) \times (\text{A.4})$ and integrating over all space. We have

© The Author 2015
S. Hanasoge, *Imaging Convection and Magnetism in the Sun*, SpringerBriefs in Mathematics, DOI 10.1007/978-3-319-27330-3

$$\int_{\odot} d\mathbf{x}\, G_{qi}^{\dagger}(\mathbf{x},\mathbf{x}_B)\, [(\mathscr{H} - 2i\omega\mathbf{v}\cdot\boldsymbol{\nabla})\mathbf{G}(\mathbf{x},\mathbf{x}_A)]_{ip} - G_{pi}(\mathbf{x},\mathbf{x}_A)\, [(\mathscr{H} + 2i\omega\mathbf{v}\cdot\boldsymbol{\nabla})\mathbf{G}^{\dagger}(\mathbf{x},\mathbf{x}_B)]_{iq}$$

$$= \int_{\odot} d\mathbf{x}\, G_{qi}^{\dagger}(\mathbf{x},\mathbf{x}_B)\, \delta_{ip}\, \delta(\mathbf{x}-\mathbf{x}_A) - G_{pi}(\mathbf{x},\mathbf{x}_A)\, \delta_{iq}\, \delta(\mathbf{x}-\mathbf{x}_B). \tag{A.5}$$

Equations (A.1) and (A.2) imply the left-hand side of (A.5) is zero. Thus we arrive at the seismic reciprocity relation for helioseismic waves

$$G_{qp}^{\dagger}(\mathbf{x}_A,\mathbf{x}_B,\omega) = G_{pq}(\mathbf{x}_B,\mathbf{x}_A,\omega), \tag{A.6}$$

where $G^{\dagger}$ is Green's function for an identical wave operator, except with flows reversed in sign. The adjoint operator $\mathscr{L}^{\dagger}$ is therefore

$$\mathscr{L}^{\dagger}\boldsymbol{\xi} = -\omega^2\rho\boldsymbol{\xi} - i\omega\rho\Gamma\boldsymbol{\xi} + 2i\omega\rho\mathbf{v}\cdot\boldsymbol{\nabla}\boldsymbol{\xi} - \boldsymbol{\nabla}(c^2\rho\boldsymbol{\nabla}\cdot\boldsymbol{\xi} + \boldsymbol{\xi}\cdot\boldsymbol{\nabla}p) - \boldsymbol{\nabla}\cdot(\rho\boldsymbol{\xi})\mathbf{g}$$
$$- [\boldsymbol{\nabla}\times\mathbf{B}\times\boldsymbol{\nabla}\times(\boldsymbol{\xi}\times\mathbf{B}) + \{\boldsymbol{\nabla}\times[\boldsymbol{\nabla}\times(\boldsymbol{\xi}\times\mathbf{B})]\}\times\mathbf{B}]. \tag{A.7}$$

A.2 Magnetic field kernels

The perturbed magnetic operator is described by

$$\delta\mathscr{L}\boldsymbol{\xi} = -(\boldsymbol{\nabla}\times\delta\mathbf{B})\times[\boldsymbol{\nabla}\times(\boldsymbol{\xi}\times\mathbf{B})] - \{\boldsymbol{\nabla}\times[\boldsymbol{\nabla}\times(\boldsymbol{\xi}\times\delta\mathbf{B})]\}\times\mathbf{B}$$
$$- (\boldsymbol{\nabla}\times\mathbf{B})\times[\boldsymbol{\nabla}\times(\boldsymbol{\xi}\times\delta\mathbf{B})] - \{\boldsymbol{\nabla}\times[\boldsymbol{\nabla}\times(\boldsymbol{\xi}\times\mathbf{B})]\}\times\delta\mathbf{B}. \tag{A.8}$$

The variation in the misfit is given by

$$\delta\mathscr{I}_1 = \frac{1}{T}\sum_{\alpha,\beta}\int_{\odot} d\mathbf{x}\int d\omega\, \boldsymbol{\Phi}^{\dagger}{}_{\alpha\beta}\cdot\bigg\{(\boldsymbol{\nabla}\times\delta\mathbf{B})\times[\boldsymbol{\nabla}\times(\boldsymbol{\Phi}_{\alpha}\times\mathbf{B})]$$
$$+ \{\boldsymbol{\nabla}\times[\boldsymbol{\nabla}\times(\boldsymbol{\Phi}_{\alpha}\times\delta\mathbf{B})]\}\times\mathbf{B}$$
$$+ (\boldsymbol{\nabla}\times\mathbf{B})\times[\boldsymbol{\nabla}\times(\boldsymbol{\Phi}_{\alpha}\times\delta\mathbf{B})] + \{\boldsymbol{\nabla}\times[\boldsymbol{\nabla}\times(\boldsymbol{\Phi}_{\alpha}\times\mathbf{B})]\}\times\delta\mathbf{B}\bigg\}. \tag{A.9}$$

In order to free the $\delta\mathbf{B}$ from the confines of the differential curl operator, we make use of the following vector identities,

$$\mathbf{a}\cdot(\mathbf{b}\times\mathbf{c}) = \mathbf{c}\cdot(\mathbf{a}\times\mathbf{b}) = \mathbf{b}\cdot(\mathbf{c}\times\mathbf{a}) \tag{A.10}$$

$$\mathbf{a}\cdot\boldsymbol{\nabla}\times\mathbf{b} = \mathbf{b}\cdot\boldsymbol{\nabla}\times\mathbf{a} - \boldsymbol{\nabla}\cdot(\mathbf{a}\times\mathbf{b}), \tag{A.11}$$

and the fact that

$$\int_{\odot} d\mathbf{x}\, \boldsymbol{\nabla}\cdot(\mathbf{a}\times\mathbf{b}) = 0, \tag{A.12}$$

due to the homogeneous upper boundary conditions we employ. Taking the first term, we have

$$\boldsymbol{\Phi}^{\dagger}{}_{\alpha\beta} \cdot \left[(\boldsymbol{\nabla} \times \delta\mathbf{B}) \times \boldsymbol{\nabla} \times (\boldsymbol{\Phi}_\alpha \times \mathbf{B}) \right] = (\boldsymbol{\nabla} \times \delta\mathbf{B}) \cdot \left\{ [\boldsymbol{\nabla} \times (\boldsymbol{\Phi}_\alpha \times \mathbf{B})] \times \boldsymbol{\Phi}^{\dagger}{}_{\alpha\beta} \right\},$$

$$\int_{\odot} d\mathbf{x}\, (\boldsymbol{\nabla} \times \delta\mathbf{B}) \cdot \left[\boldsymbol{\nabla} \times (\boldsymbol{\Phi}_\alpha \times \mathbf{B}) \times \boldsymbol{\Phi}^{\dagger}{}_{\alpha\beta} \right] = \int_{\odot} d\mathbf{x}\, \delta\mathbf{B} \cdot \left\{ \boldsymbol{\nabla} \times [\boldsymbol{\nabla} \times (\boldsymbol{\Phi}_\alpha \times \mathbf{B}) \times \boldsymbol{\Phi}^{\dagger}{}_{\alpha\beta}] \right\},$$

and now the second,

$$\boldsymbol{\Phi}^{\dagger}{}_{\alpha\beta} \cdot \left[\{\boldsymbol{\nabla} \times [\boldsymbol{\nabla} \times (\boldsymbol{\Phi}_\alpha \times \delta\mathbf{B})]\} \times \mathbf{B} \right] = \{\boldsymbol{\nabla} \times [\boldsymbol{\nabla} \times (\boldsymbol{\Phi}_\alpha \times \delta\mathbf{B})]\} \cdot (\mathbf{B} \times \boldsymbol{\Phi}^{\dagger}{}_{\alpha\beta}),$$

$$\int_{\odot} d\mathbf{x}\, \{\boldsymbol{\nabla} \times [\boldsymbol{\nabla} \times (\boldsymbol{\Phi}_\alpha \times \delta\mathbf{B})]\} \cdot (\mathbf{B} \times \boldsymbol{\Phi}^{\dagger}{}_{\alpha\beta}) = \int_{\odot} d\mathbf{x}\, [\boldsymbol{\nabla} \times (\boldsymbol{\Phi}_\alpha \times \delta\mathbf{B})] \cdot \boldsymbol{\nabla} \times (\mathbf{B} \times \boldsymbol{\Phi}^{\dagger}{}_{\alpha\beta}),$$

$$\int_{\odot} d\mathbf{x}\, [\boldsymbol{\nabla} \times (\boldsymbol{\Phi}_\alpha \times \delta\mathbf{B})] \cdot \boldsymbol{\nabla} \times (\mathbf{B} \times \boldsymbol{\Phi}^{\dagger}{}_{\alpha\beta}) = \int_{\odot} d\mathbf{x}\, (\boldsymbol{\Phi}_\alpha \times \delta\mathbf{B}) \cdot \boldsymbol{\nabla} \times [\boldsymbol{\nabla} \times (\mathbf{B} \times \boldsymbol{\Phi}^{\dagger}{}_{\alpha\beta})],$$

$$(\boldsymbol{\Phi}_\alpha \times \delta\mathbf{B}) \cdot \boldsymbol{\nabla} \times [\boldsymbol{\nabla} \times (\mathbf{B} \times \boldsymbol{\Phi}^{\dagger}{}_{\alpha\beta})] = \delta\mathbf{B} \cdot \left\{ \boldsymbol{\nabla} \times [\boldsymbol{\nabla} \times (\mathbf{B} \times \boldsymbol{\Phi}^{\dagger}{}_{\alpha\beta})] \times \boldsymbol{\Phi}_\alpha \right\},$$

followed by the third

$$\boldsymbol{\Phi}^{\dagger}{}_{\alpha\beta} \cdot \left[(\boldsymbol{\nabla} \times \mathbf{B}) \times \boldsymbol{\nabla} \times (\boldsymbol{\Phi}_\alpha \times \delta\mathbf{B}) \right] = \boldsymbol{\nabla} \times (\boldsymbol{\Phi}_\alpha \times \delta\mathbf{B}) \cdot \left[\boldsymbol{\Phi}^{\dagger}{}_{\alpha\beta} \times (\boldsymbol{\nabla} \times \mathbf{B}) \right],$$

$$\int_{\odot} d\mathbf{x}\, \boldsymbol{\nabla} \times (\boldsymbol{\Phi}_\alpha \times \delta\mathbf{B}) \cdot \left[\boldsymbol{\Phi}^{\dagger}{}_{\alpha\beta} \times (\boldsymbol{\nabla} \times \mathbf{B}) \right] = \int_{\odot} d\mathbf{x}\, (\boldsymbol{\Phi}_\alpha \times \delta\mathbf{B}) \cdot \left\{ \boldsymbol{\nabla} \times [\boldsymbol{\Phi}^{\dagger}{}_{\alpha\beta} \times (\boldsymbol{\nabla} \times \mathbf{B})] \right\},$$

$$(\boldsymbol{\Phi}_\alpha \times \delta\mathbf{B}) \cdot \left\{ \boldsymbol{\nabla} \times [\boldsymbol{\Phi}^{\dagger}{}_{\alpha\beta} \times (\boldsymbol{\nabla} \times \mathbf{B})] \right\} = \delta\mathbf{B} \cdot \left\{ \boldsymbol{\nabla} \times [\boldsymbol{\Phi}^{\dagger}{}_{\alpha\beta} \times (\boldsymbol{\nabla} \times \mathbf{B})] \times \boldsymbol{\Phi}_\alpha \right\},$$

and finally, the simplest of them all

$$\boldsymbol{\Phi}^{\dagger}{}_{\alpha\beta} \cdot \left\{ \boldsymbol{\nabla} \times [\boldsymbol{\nabla} \times (\boldsymbol{\Phi}_\alpha \times \mathbf{B})] \times \delta\mathbf{B} \right\} = \delta\mathbf{B} \cdot \left\{ \boldsymbol{\Phi}^{\dagger}{}_{\alpha\beta} \times \boldsymbol{\nabla} \times [\boldsymbol{\nabla} \times (\boldsymbol{\Phi}_\alpha \times \mathbf{B})] \right\}.$$

There are other terms that arise from perturbing the equilibrium equation (3.67). These are

$$-\frac{1}{T} \int_{\odot} d\mathbf{x}\, (\boldsymbol{\nabla} \cdot \boldsymbol{\Phi}^{\dagger}{}_{\alpha\beta}) \, \boldsymbol{\Phi}_\alpha \cdot [(\boldsymbol{\nabla} \times \delta\mathbf{B}) \times \mathbf{B} + (\boldsymbol{\nabla} \times \mathbf{B}) \times \delta\mathbf{B}].$$

Expanding on the first,

$$(\boldsymbol{\nabla} \cdot \boldsymbol{\Phi}^{\dagger}{}_{\alpha\beta}) \, \boldsymbol{\Phi}_\alpha \cdot (\boldsymbol{\nabla} \times \delta\mathbf{B}) \times \mathbf{B} = (\boldsymbol{\nabla} \times \delta\mathbf{B}) \cdot [\mathbf{B} \times (\boldsymbol{\Phi}_\alpha \boldsymbol{\nabla} \cdot \boldsymbol{\Phi}^{\dagger}{}_{\alpha\beta})],$$

$$\int_{\odot} d\mathbf{x}\, (\boldsymbol{\nabla} \times \delta\mathbf{B}) \cdot [\mathbf{B} \times (\boldsymbol{\Phi}_\alpha \boldsymbol{\nabla} \cdot \boldsymbol{\Phi}^{\dagger}{}_{\alpha\beta})] = \int_{\odot} d\mathbf{x}\, \delta\mathbf{B} \cdot \{\boldsymbol{\nabla} \times [\mathbf{B} \times (\boldsymbol{\Phi}_\alpha \boldsymbol{\nabla} \cdot \boldsymbol{\Phi}^{\dagger}{}_{\alpha\beta})]\},$$

and the second may be manipulated so

$$(\boldsymbol{\nabla} \cdot \boldsymbol{\Phi}^{\dagger}{}_{\alpha\beta}) \, \boldsymbol{\Phi}_\alpha \cdot (\boldsymbol{\nabla} \times \mathbf{B}) \times \delta\mathbf{B} = \delta\mathbf{B} \cdot [\boldsymbol{\nabla} \cdot \boldsymbol{\Phi}^{\dagger}{}_{\alpha\beta} \boldsymbol{\Phi}_\alpha \times (\boldsymbol{\nabla} \times \mathbf{B})]. \tag{A.13}$$

A.3 Units of Wavefields, Conventions, and Definitions

We apply the following Fourier transform convention

$$\int_{-\infty}^{\infty} dt\, e^{i\omega t}\, g(t) = g(\omega),$$ (A.14)

$$\int_{-\infty}^{\infty} dt\, e^{i\omega t} = 2\pi\, \delta(\omega),$$ (A.15)

$$\frac{1}{2\pi}\int_{-\infty}^{\infty} d\omega\, e^{-i\omega t}\, g(\omega) = g(t),$$ (A.16)

$$\int_{-\infty}^{\infty} d\omega\, e^{-i\omega t} = 2\pi\, \delta(t),$$ (A.17)

where the Fourier transform pair $g(t), g(\omega)$ are written similarly for convenience. The equivalence between cross correlations and convolutions in the Fourier and temporal domain are written so

$$h(t) = \int_{-\infty}^{\infty} dt'\, f(t')\, g(t+t') \Longleftrightarrow h(\omega) = f^*(\omega)\, g(\omega),$$ (A.18)

$$h(t) = \int_{-\infty}^{\infty} dt'\, f(t')\, g(t-t') \Longleftrightarrow h(\omega) = f(\omega)\, g(\omega).$$ (A.19)

The following relationship also holds (for real functions $f(t), g(t)$)

$$\int_{-\infty}^{\infty} dt\, f(t)\, g(t) = \frac{1}{2\pi}\int_{-\infty}^{\infty} d\omega\, f^*(\omega)\, g(\omega) = \frac{1}{2\pi}\int_{-\infty}^{\infty} d\omega\, f(\omega)\, g^*(\omega).$$ (A.20)

We now describe the physical units of various quantities (indicated by square brackets around a given variable)

- $[\delta(\mathbf{x})] \equiv \mathrm{Mm}^{-3}$ $(\int_\odot d\mathbf{x}\, \delta(\mathbf{x}) = 1)$
- $[\delta(t)] \equiv \mathrm{s}^{-1}$ $(\int dt\, \delta(t) = 1)$
- $[\mathscr{L}] \equiv \mathrm{g}\cdot\mathrm{Mm}^{-3}\cdot\mathrm{s}^{-2}$ $(\mathscr{L} \sim \rho\omega^2)$
- $[\mathbf{G}] \equiv \mathrm{s}\cdot\mathrm{g}^{-1}$ $[\mathscr{L}\mathbf{G} = \delta(\mathbf{x}-\mathbf{x}')\delta(t-t')]$
- $[\mathbf{S}] \equiv \mathrm{g}\cdot\mathrm{Mm}^{-3}\cdot\mathrm{Mm}\cdot\mathrm{s}^{-2}$ $[\mathscr{L}\boldsymbol{\xi}(\mathbf{x},t) = \mathbf{S}(\mathbf{x},t)]$
- $[\mathscr{G}] \equiv \mathrm{g}^{-1}$ $[\int_\odot d\mathbf{x}'\, dt'\, \mathscr{G}(\mathbf{x},\mathbf{x}',t-t')\cdot\mathbf{S}(\mathbf{x}',t') = \phi(\mathbf{x},t)]$
- $[\mathscr{F}] \equiv \mathrm{Mm}^{-3}\,\mathrm{s}^{-2}$ $[\mathscr{G}_j(\mathbf{x},\mathbf{x}'',t) = \int dt'\, d\mathbf{x}'\, \mathscr{F}(\mathbf{x}',t')\, l_i\, G_{ij}(\mathbf{x}-\mathbf{x}',\mathbf{x}'',t-t')]$
- $[\mathscr{P}] \equiv \mathrm{g}^2\cdot\mathrm{Mm}^{-3}\cdot\mathrm{Mm}^2\cdot\mathrm{s}^{-3}$ $[\mathscr{P}_{ij}(\mathbf{x},\omega)\delta(\mathbf{x}-\mathbf{x}') = \langle S_i(\mathbf{x},\omega)S_j^*(\mathbf{x}',\omega)\rangle]$
- $[\boldsymbol{\eta}] \equiv \mathrm{g}\cdot\mathrm{Mm}^{-1}\cdot\mathrm{s}^{-2}$ $[= \int dt'\, \mathscr{G}(\mathbf{x},\mathbf{x}_\alpha,t-t')\cdot\mathscr{P}(\mathbf{x},t')]$
- $[\mathscr{M}_i] \equiv \mathrm{Mm}^{-5}\cdot\mathrm{s}^2$ $[= \int dt'\, l_i b_q\, W_{\alpha\beta}(t'+t)\, \mathscr{F}(\mathbf{x}_\beta - \mathbf{x}',t')]$
- $[\boldsymbol{\Phi}] \equiv \mathrm{Mm}^2$ $[= \int dt'\, d\mathbf{x}'\, \mathbf{G}(\mathbf{x},\mathbf{x}',t-t')\cdot\boldsymbol{\eta}(\mathbf{x}',t')]$
- $[\boldsymbol{\Phi}^\dagger] \equiv \mathrm{g}^{-1}\cdot\mathrm{Mm}^{-2}\cdot\mathrm{s}^4$ $[= \int dt'\, d\mathbf{x}'\, \mathbf{G}(\mathbf{x},\mathbf{x}',t-t')\cdot\mathscr{M}(\mathbf{x}',t')]$
- $[\mathbf{K}_v] \equiv \mathrm{Mm}^{-4}\cdot\mathrm{s}^3$ $[= \frac{1}{T}\int dt\, \rho[\boldsymbol{\nabla}\partial_t\boldsymbol{\Phi}(t)]\cdot\boldsymbol{\Phi}^\dagger(-t)]$

We use equation (4) from Gizon (2004) in order to define the weight function $W_{\alpha\beta}(t)$ for the differential flow measurement

$$W_{\alpha\beta}(t) = -\dot{\mathscr{C}}_{\alpha\beta}(t) \frac{f(t)+f(-t)}{\Delta t \sum_{t'} f(t') \left[\dot{\mathscr{C}}_{\alpha\beta}(t')\right]^2}, \tag{A.21}$$

where Δt is the temporal rate at which the cross correlations are sampled, $f(t)$ is a window, and the difference travel time $\delta\tau$ is given by

$$\delta\tau = \int dt \, W_{\alpha\beta}(t) \, \delta\mathscr{C}_{\alpha\beta}(t). \tag{A.22}$$

Note that since we compute difference travel times, $W_{\alpha\beta}$ is an odd function of time whose Fourier transform is therefore purely imaginary. For the sound-speed kernel, we measure mean travel times, defined as

$$W_{\alpha\beta}(t) = -\frac{1}{2} \dot{\mathscr{C}}_{\alpha\beta}(t) \frac{f(t)-f(-t)}{\Delta t \sum_{t'} f(t') \left[\dot{\mathscr{C}}_{\alpha\beta}(t')\right]^2}. \tag{A.23}$$

A.4 Validation

We perform validation tests in order to test the quality of computed kernels and limit cross correlations.

A.4.1 Classical-tomographic sound-speed kernel

As a simple test, we compute a single-source sound-speed kernel between a pair of points located 15 Mm apart. The source point is forced with the function shown in the upper-most panel of Figure A.1; this calculation forms the forward wavefield. The seismogram at the receiver 15 Mm away is shown in the middle panel where the dot-dash lines denote the temporal window applied to isolate the first arrival. The adjoint source, the time-reversed windowed seismogram, is applied at the receiver. The kernel is subsequently calculated according to equation (3.66) and is shown in Figure A.2.

Consider the travel time of a ray propagating along path $\mathscr{R}$

$$\tau = \int_{\mathscr{R}} \frac{ds}{c}, \tag{A.24}$$

where s is length measured along raypath $\mathscr{R}$. Fermat assures us that the raypath is invariant under small perturbations of sound speed. Therefore the perturbation in travel time due to spatially constant $\delta c/c$ is given by

$$\delta \tau = -\frac{\delta c}{c} \int_{\mathscr{R}} \frac{ds}{c} = -\tau \frac{\delta c}{c}. \tag{A.25}$$

The sound-speed kernel must therefore satisfy (having divided out $\Delta \tau$),

$$\delta \tau = \int_{\odot} d\mathbf{x} \, \frac{\delta c^2}{c^2} K_{c^2}(\mathbf{x}) \approx -\tau \frac{\delta c}{c}, \tag{A.26}$$

or

$$\int_{\odot} d\mathbf{x} \, K_{c^2}(\mathbf{x}) \approx -\frac{\tau}{2}. \tag{A.27}$$

We find the integral of the kernel to be -192.88 s, which compares well with half the travel time, -190 s.

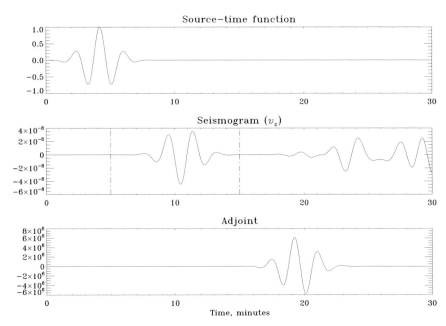

Fig. A.1 Source-time function, receiver seismogram, and adjoint source involved in the computation of a single-source sound-speed kernel (e.g., Tromp et al 2005). Vertical and horizontal cuts are shown. The dot-dash lines in the seismogram show the temporal window applied to isolate the first arrival.

A.4.2 Cross correlations

The filtered cross correlation for a translationally invariant background model may be written as

$$\mathscr{C}(\boldsymbol{\Delta}, \omega) = \int d\mathbf{k} \, |\phi(\mathbf{k}, \omega) \mathscr{F}(\mathbf{k}, \omega)|^2 \, e^{i\mathbf{k} \cdot \boldsymbol{\Delta}}, \tag{A.28}$$

where Δ is the vector connecting two observation points. Thus we may estimate the cross correlation between a given pair of points by inverse Fourier transforming the power spectrum. In Figures 3.1 and 3.4, we compare computed and spectrally estimated cross correlations and find some differences that likely arise from the finite-size of the horizontal domain and absorption boundary conditions that dissipate high-group-speed (low-frequency) waves which reach boundaries first.

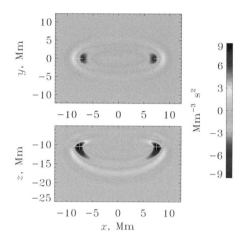

Fig. A.2 A single-source sound-speed kernel (e.g., Gizon and Birch 2002; Tromp et al 2005). Vertical and horizontal cuts are shown. The symbols denote source (left) and receiver positions. The integral of the kernel is -192.88 s, compared to a half wave travel time of -190 s.

A.4.3 Flow-kernel Integral

We introduce a spatially uniform 0.1 km/s x-directed flow (i.e., everywhere in the domain) and the corresponding travel-time shift using equations (A.21) and (3.24) is -9.6 s. The change in cross correlation for this background model is displayed in Figure A.3. The flow-kernel integral is

$$\delta\tau = v_x \int_\odot d\mathbf{x}\, K_{v_x}(\mathbf{x}) = 0.1 \int_\odot d\mathbf{x}\, K_{v_x}(\mathbf{x}) = -10.1\,\text{s}. \qquad (A.29)$$

As expected, integrals of K_{v_y} and K_{v_z} are zero (Birch and Gizon 2007).

A.4.4 Integral of the multiple-source sound-speed kernel

We introduce a spatially uniform 1% perturbation to c^2 (i.e., everywhere in the domain) and the corresponding travel-time shift computed using equations (A.23) and (3.24) is -1.98 s. The change in cross correlation for this slightly altered background model is displayed in Figure A.4.

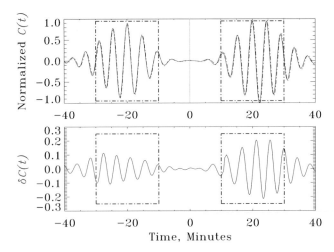

Fig. A.3 The upper panel displays cross correlations between a point pair 10 Mm apart, corresponding to background models with no flow (solid line) and a constant x-directed flow of magnitude 0.1 km/s. The related travel-time shift, computed using (A.21) and (3.24), is -9.6 s, which implies a kernel integral of -96 s.

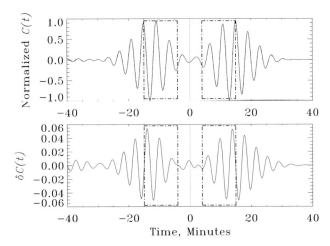

Fig. A.4 The upper panel displays cross correlations between a point pair 10 Mm apart, corresponding to background models with sound-speed distributions of c^2 and $1.01c^2$. The two cross correlations fall almost on top of other and their difference, only visible on the lower panel, is on the order of a few percent. The related travel-time shift, computed using (A.23) and (3.24), is -1.98 s, implying a kernel integral of -198 s.

The expected travel-time shift for such a perturbation is given by the following integral

$$\delta \tau = \int_{\odot} d\mathbf{x} \, K_{c^2}(\mathbf{x}) \frac{\delta c^2}{c^2} = 0.01 \int_{\odot} d\mathbf{x} \, K_{c^2}(\mathbf{x}) = -1.75 \, \mathrm{s}, \qquad (A.30)$$

where the kernel is displayed in Figure 3.5

A.5 Adjoint source

We use equation (4) from Gizon and Birch (2004) in order to define the weight function $W_i(t)$ for the travel-time measurement

$$W_i(t) = -\dot{\mathscr{C}}_i^{\mathrm{p}}(t) \frac{f(t)}{\Delta t \sum_{t'} f(t') \left[\dot{\mathscr{C}}_i^{\mathrm{p}}(t')\right]^2}, \qquad (A.31)$$

where $\mathscr{C}^{\mathrm{p}}$ is the predicted waveform (cross correlation), Δt is the temporal rate at which the waveform is sampled, $f(t)$ is a window, and the travel-time shift $\Delta \tau$ is given by

$$\Delta \tau_i = \int dt \, W_i(t) \, (\mathscr{C}_i^{\mathrm{p}} - \mathscr{C}_i^{\mathrm{o}}). \qquad (A.32)$$

The adjoint source is given by

$$f^{\dagger}(\mathbf{x}, t) = \sum_i \Delta \tau_i W_i(-t) \, \delta(\mathbf{x} - \mathbf{x}_i), \qquad (A.33)$$

where $\mathbf{x}_i$ is the receiver (slave) and the summation is over all receivers.

A.6 Steepest descent, Conjugate gradient, and L-BFGS

In all the methods described here, the model is updated thus, $\mathbf{m}^{k+1} = \mathbf{m}^k + \varepsilon \mathbf{p}^k$, where ε is obtained through a line search, i.e., ε that minimizes $\chi(\mathbf{m}^k + \varepsilon \mathbf{p}^k)$. Given the smoothed gradient at iteration k, $\mathbf{g}^k$. The steepest descent update is simply $\mathbf{p}^k = -\mathbf{g}^k$. The conjugate gradient update is given by

$$\mathbf{p}^k = -\mathbf{g}^k + \beta^k \mathbf{p}^{k-1}, \qquad \beta^k = \frac{\mathbf{g}^k \cdot (\mathbf{g}^k - \mathbf{g}^{k-1})}{\mathbf{g}^k \cdot \mathbf{g}^k}, \qquad (A.34)$$

and because there is a dependence on $\mathbf{p}^{k-1}$, the first iteration cannot also be performed by conjugate gradient.

The limited-memory BFGS update at iteration N is obtained by manipulating the prior m gradients and models. The limited-memory aspect of this is accomplished by sweeping forward and reverse through prior gradients.

$$k = N \quad \mathbf{h} = \mathbf{g}^k$$

For k = N − 1, N − 2, ..., N − m

$$\alpha^k = \frac{(\mathbf{m}^k - \mathbf{m}^{k-1}) \cdot \mathbf{h}}{(\mathbf{m}^k - \mathbf{m}^{k-1}) \cdot (\mathbf{g}^k - \mathbf{g}^{k-1})}$$

$$\mathbf{h} = \mathbf{h} - \alpha^k(\mathbf{g}^k - \mathbf{g}^{k-1}), \tag{A.35}$$

For k = N − m, N − m + 1, ..., N − 1

$$\alpha^k = \alpha^k - \frac{(\mathbf{g}^k - \mathbf{g}^{k-1}) \cdot \mathbf{h}}{(\mathbf{m}^k - \mathbf{m}^{k-1}) \cdot (\mathbf{g}^k - \mathbf{g}^{k-1})}$$

$$\mathbf{h} = \mathbf{h} + \alpha^k(\mathbf{m}^k - \mathbf{m}^{k-1}) \tag{A.36}$$

The update is given by $\mathbf{p}^N = -\mathbf{h}$. The rule of thumb is to use between 3 and 7 prior gradients to construct the update, i.e., $3 \le m \le 7$ in equations (A.35) and (A.36).

References

Adams MF (2007) Algebraic multigrid techniques for strongly indefinite linear systems from direct frequency response analysis in solid mechanics. Comput Mech 39(4):497–507

Appelö D, Hagstrom T, Kreiss G (2006) Perfectly matched layers for hyperbolic systems: general formulation, well-posedness, and stability. SIAM J Appl Math 67(1):1–23

Balbus SA (2009) A simple model for solar isorotational contours. Mon Not R Astron Soc 395:2056–2064. DOI 10.1111/j.1365-2966.2009.14469.x. 0809.2883

Balbus SA, Schaan E (2012) The stability of stratified, rotating systems and the generation of vorticity in the sun. Mon Not R Astron Soc 426:1546–1557. DOI 10.1111/j.1365-2966.2012.21729.x. 1207.3810

Baldner CS, Schou J (2012) Effects of asymmetric flows in solar convection on oscillation modes. Astrophys J Lett 760:L1. DOI 10.1088/2041-8205/760/1/L1. 1210.1583

Bamberger A, Chavent G, Hemons C, Lailly P (1982) Inversion of normal incidence seismograms. Geophysics 47:757–770

Bécache E, Fauqueux S, Joly P (2003) Stability of perfectly matched layers, group velocities and anisotropic waves. J Comput Phys 188(2):399–433

Beck JG, Duvall TL Jr (2001) Time-distance study of supergranulation. In: Wilson A, Scientific coordination by Pallè PL (eds) Proceedings of the SOHO 10/GONG 2000 workshop: helio- and asteroseismology at the dawn of the millennium, Santa Cruz de Tenerife, 2-6 October 2000. ESA SP-464. ESA, Noordwijk, pp 577–581. ISBN 92-9092-697-X. http://adsabs.harvard.edu/abs/2001ESASP.464..577B

Beliën AJC, Botchev MA, Goedbloed JP, van der Holst B, Keppens R (2002) FINESSE: axisymmetric MHD equilibria with flow. J Comput Phys 182(1): 91–117. http://www.sciencedirect.com/science/article/B6WHY-481662C-5/1/3cfcd06c0b8464d381f01ee3cc1d73bd

Belousov SL (1962) Tables of normalized associated Legendre polynomials, vol 18. Pergamon Press, New York

Berenger JP (1994) A perfectly matched layer for the absorption of electromagnetic waves. J Comput Phys 114(2):185–200. DOI 10.1006/jcph.

© The Author 2015
S. Hanasoge, *Imaging Convection and Magnetism in the Sun*, SpringerBriefs in Mathematics, DOI 10.1007/978-3-319-27330-3

1994.1159. http://www.sciencedirect.com/science/article/B6WHY-45P0TJR-1P/ 2/9aa5d47e587908f2e75691914c79ac80

Bérenger JP (2002) Application of the CFS PML to the absorption of evanescent waves in waveguides. IEEE Microwave Wireless Compon Lett 12(6):218–220. DOI 10.1109/LMWC.2002.1010000

Berland J, Bogey C, Bailly C (2006) Low-dissipation and low-dispersion fourth-order Runge–Kutta algorithm. Comput Fluids 35(10):1459

Bewley TR, Moin P, Temam R (2001) DNS-based predictive control of turbulence: an optimal benchmark for feedback algorithms. J Fluid Mech 447:179–225

Bhattacharya J, Hanasoge S, Antia HM (2015) Frequency shifts of resonant modes of the sun due to near-surface convective scattering. Astrophys J 806:246. DOI 10.1088/0004-637X/806/2/246. 1505.04048

Birch A, Duvall TL, Hanasoge S (2007) Local helioseismology of supergranulation. In: American astronomical society meeting abstracts #210. Bulletin of the American Astronomical Society, vol 39, p 160

Birch AC, Gizon L (2007) Linear sensitivity of helioseismic travel times to local flows. Astron Nachr 328:228. DOI 10.1002/asna.200610724

Birch AC, Kosovichev AG, Price GH, Schlottmann RB (2001) The accuracy of the born and ray approximations in time-distance helioseismology. Astrophys J Lett 561:L229–L232. DOI 10.1086/324766

Birch AC, Kosovichev AG, Duvall TL Jr (2004) Sensitivity of acoustic wave travel times to sound-speed perturbations in the solar interior. Astrophys J 608:580–600. DOI 10.1086/386361

Birch AC, Braun DC, Hanasoge SM, Cameron R (2009) Surface-focused seismic holography of sunspots: II. Expectations from numerical simulations using sound-speed perturbations. Sol Phys 254:17–27. DOI 10.1007/s11207-008-9282-9

Bogdan TJ, Hindman BW, Cally PS, Charbonneau P (1996) Absorption of p-modes by slender magnetic flux tubes and p-mode lifetimes. Astrophys J 465:406. DOI 10.1086/177429

Böhm-Vitense E (1958) Über die wasserstoffkonvektionszone in sternen verschiedener effektivtemperaturen und leuchtkräfte. mit 5 textabbildungen. Z Astrophys 46:108

Borrero JM, Tomczyk S, Norton A, Darnell T, Schou J, Scherrer P, Bush R, Liu Y (2007) Magnetic field vector retrieval with the helioseismic and magnetic imager. Sol Phys 240:177–196. DOI 10.1007/s11207-006-0219-x. astro-ph/0611565

Bozdağ E, Trampert J, Tromp J (2011) Misfit functions for full waveform inversion based on instantaneous phase and envelope measurements. Geophys J Int 185:845–870. DOI 10.1111/j.1365-246X.2011.04970.x

Brandenburg A (2005) Distributed versus tachocline dynamos. arXiv Astrophysics e-prints arXiv:astro-ph/0512638

Braun DC, Birch AC (2008) Prospects for the detection of the deep solar meridional circulation. Astrophys J Lett 689:L161–L165. DOI 10.1086/595884. 0810.0284

Bray RJ, Loughhead RE (1974) The solar chromosphere. The international astrophysics series. Chapman and Hall, London

Bunks C, Saleck FM, Zaleski S, Chavent G (1995) Multiscale seismic waveform inversion. Geophysics 60:1457–1473

Cameron R, Gizon L, Daiffallah K (2007) SLiM: a code for the simulation of wave propagation through an inhomogeneous, magnetised solar atmosphere. Astron Nachr 328:313. DOI 10.1002/asna.200610736

Cameron R, Gizon L, Duvall TL Jr (2008) Helioseismology of sunspots: confronting observations with three-dimensional MHD simulations of wave propagation. Sol Phys 251:291–308. DOI 10.1007/s11207-008-9148-1. 0802.1603

Cameron R, Gizon L, Schunker H, Pietarila A (2010) Constructing semi-empirical sunspot models for helioseismology. arXiv e-prints 1003.0528

Candès EJ, Romberg J, Tao T (2006) Robust uncertainty principles: exact signal reconstruction from highly incomplete frequency information. IEEE Trans Inf Theory 52(2):489–509

Christensen-Dalsgaard J (2003) Lecture notes on stellar oscillations, 5th edn. http://astro.phys.au.dk/~jcd/oscilnotes/

Christensen–Dalsgaard J (2003) Lecture notes on stellar oscillations, 5th edn. http://astro.phys.au.dk/~jcd/oscilnotes/

Christensen-Dalsgaard J (2008) Lecture notes on stellar structure and evolution, 6th edn. http://astro.phys.au.dk/~jcd/evolnotes/LN_stellar_structure.pdf

Christensen-Dalsgaard J, Dappen W, Ajukov SV, Anderson ER, Antia HM, Basu S, Baturin VA, Berthomieu G, Chaboyer B, Chitre SM, Cox AN, Demarque P, Donatowicz J, Dziembowski WA, Gabriel M, Gough DO, Guenther DB, Guzik JA, Harvey JW, Hill F, Houdek G, Iglesias CA, Kosovichev AG, Leibacher JW, Morel P, Proffitt CR, Provost J, Reiter J, Rhodes EJ Jr, Rogers FJ, Roxburgh IW, Thompson MJ, Ulrich RK (1996) The current state of solar modeling. Science 272:1286

Collino F, Monk P (1998) Optimizing the perfectly matched layer. Comput Methods Appl Mech Eng 164:157–171

Colonius T (2004) Modeling artificial boundary conditions for compressible flow. Annu Rev Fluid Mech 36:315–345

Cooley JW, Tukey JW (1965) An algorithm for the machine calculation of complex fourier series. Math Comput 19(90):297–301

Dahlen FA, Baig AM (2002) Fréchet kernels for body-wave amplitudes. Geophys J Int 150:440–466. DOI 10.1046/j.1365-246X.2002.01718.x

DeGrave K, Jackiewicz J, Rempel M (2014) Validating time-distance helioseismology with realistic quiet-sun simulations. Astrophys J 788:127. DOI 10.1088/0004-637X/788/2/127. 1404.4645

Deubner FL (1975) Observations of low wavenumber nonradial eigenmodes of the sun. Astron Astrophys 44:371–375

Dombroski DE, Birch AC, Braun DC, Hanasoge SM (2013) Testing helioseismic-holography inversions for supergranular flows using synthetic data. Sol Phys 282:361–378. DOI 10.1007/s11207-012-0189-0. 1211.6886

Duvall TL, Hanasoge SM (2013) Subsurface supergranular vertical flows as measured using large distance separations in time-distance helioseismology. Sol Phys 287:71–83. DOI 10.1007/s11207-012-0010-0. 1207.6075

Duvall TL, Hanasoge SM, Chakraborty S (2014) Additional evidence support-
ing a model of shallow, high-speed supergranulation. Sol Phys 289:3421–3433.
DOI 10.1007/s11207-014-0537-3. 1404.2533

Duvall TL Jr (1982) A dispersion law for solar oscillations. Nature 300:242

Duvall TL Jr (2003) Nonaxisymmetric variations deep in the convection zone.
In: Sawaya-Lacoste H (ed) Proceedings of SOHO 12/GONG+ 2002. Local and
global helioseismology: the present and future, Big Bear Lake, 27 October-1
November 2002. ESA SP-517. ESA, Noordwijk, pp 259–262. ISBN 92-9092-
827-1. http://adsabs.harvard.edu/abs/2003ESASP.517..259D

Duvall TL Jr, Gizon L (2000) Time-distance helioseismology with f modes as a
method for measurement of near-surface flows. Sol Phys 192:177–191. DOI 10.
1023/A:1005239503637

Duvall TL Jr, Harvey JW, Pomerantz MA (1986) Latitude and depth variation of
solar rotation. Nature 321:500

Duvall TL Jr, Jefferies SM, Harvey JW, Pomerantz MA (1993) Time-distance
helioseismology. Nature 362:430–432. DOI 10.1038/362430a0

Duvall TL Jr, D'Silva S, Jefferies SM, Harvey JW, Schou J (1996) Downflows under
sunspots detected by helioseismic tomography. Nature 379:235. DOI 10.1038/
379235a0

Duvall TL Jr, Kosovichev AG, Murawski K (1998) Random damping and frequency
reduction of the solar f mode. Astrophys J Lett 505:L55. DOI 10.1086/311595

Duvall TL Jr, Birch AC, Gizon L (2006) Direct measurement of travel-time kernels
for helioseismology. Astrophys J 646:553–559, DOI 10.1086/504792

Felipe T, Khomenko E, Collados M (2010) Magneto-acoustic waves in sunspots:
first results from a new three-dimensional nonlinear magnetohydrodynamic code.
Astrophys J 719:357–377. DOI 10.1088/0004-637X/719/1/357. 1006.2998

Felipe T, Crouch AD, Birch AC (2014) Evaluation of the capability of local helioseismology to discern between monolithic and spaghetti sunspot models. Astrophys J 788:136. DOI 10.1088/0004-637X/788/2/136. 1405.0036

Festa G, Vilotte JP (2005) The Newmark scheme as velocity-stress time-staggering:
an efficient PML implementation for spectral-element simulations of elastody-
namics. Geophys J Int 161:789–812. DOI 10.1111/j.1365-246X.2005.02601.x

Fichtner A, Kennett BLN, Igel H, Bunge H (2008) Theoretical background for
continental- and global-scale full-waveform inversion in the time-frequency
domain. Geophys J Int 175:665–685. DOI 10.1111/j.1365-246X.2008.03923.x

Fichtner A, Kennett BLN, Igel H, Bunge HP (2009) Full seismic waveform tomography for upper-mantle structure in the Australasian region using adjoint methods.
Geophys J Int 179:1703–1725. DOI 10.1111/j.1365-246X.2009.04368.x

Gee LS, Jordan TH (1992) Generalized seismological data functionals. Geophys J
Int 111:363–390. DOI 10.1111/j.1365-246X.1992.tb00584.x

Giles MB, Pierce NA (2000) An introduction to the adjoint approach to design. Flow
Turbul Combust 65:393–415

Giles PM (2000) Time-distance measurements of large-scale flows in the solar convection zone. Ph.D. thesis, Stanford University, CA, USA

Giles PM, Duvall TL Jr, Scherrer PH, Bogart RS (1997) A flow of material from the suns equator to its poles. Nature 390:52

Gizon L (2003) Probing flows in the upper solar convection zone. Ph.D. thesis, Stanford University, CA, USA

Gizon L (2004) Helioseismology of time-varying flows through the solar cycle. Sol Phys 224:217–228. DOI 10.1007/s11207-005-4983-9

Gizon L, Birch AC (2002) Time-distance helioseismology: the forward problem for random distributed sources. Astrophys J 571:966–986. DOI 10.1086/340015

Gizon L, Birch AC (2004) Time-distance helioseismology: noise estimation. Astrophys J 614:472–489. DOI 10.1086/423367

Gizon L, Birch AC (2005) Local helioseismology. Living Rev Sol Phys 2:6

Gizon L, Duvall TL, Schou J (2003) Wave-like properties of solar supergranulation. Nature 421:43–44. arXiv:astro-ph/0208343

Gizon L, Hanasoge SM, Birch AC (2006) Scattering of acoustic waves by a magnetic cylinder: accuracy of the born approximation. Astrophys J 643:549–555. DOI 10.1086/502623. arXiv:0803.3839

Gizon L, Schunker H, Baldner CS, Basu S, Birch AC, Bogart RS, Braun DC, Cameron R, Duvall TL, Hanasoge SM, Jackiewicz J, Roth M, Stahn T, Thompson MJ, Zharkov S (2009) Helioseismology of sunspots: a case study of NOAA region 9787. Space Sci Rev 144:249–273. DOI 10.1007/s11214-008-9466-5

Gizon L, Birch AC, Spruit HC (2010) Local helioseismology: three-dimensional imaging of the solar interior. Annu Rev Astron Astrophys 48:289–338. DOI 10.1146/annurev-astro-082708-101722. 1001.0930

Goedbloed JPH, Poedts S (2004) Principles of magnetohydrodynamics: with applications to laboratory and astrophysical plasmas. Cambridge University Press, Cambridge. DOI 10.2277/0521626072

Haigh JD (2007) The sun and the earth's climate. Living Rev Sol Phys 4(2). http://www.livingreviews.org/lrsp-2007-2; http://adsabs.harvard.edu/abs/2007LRSP....4....2H

Hanasoge S, Birch A, Gizon L, Tromp J (2012a) Seismic probes of solar interior magnetic structure. Phys Rev Lett 109(10):101101. DOI 10.1103/PhysRevLett.109.101101. 1207.4352

Hanasoge SM (2007) Theoretical studies of wave interactions in the sun. Ph.D. thesis, Stanford University, California, USA

Hanasoge SM (2008) Seismic halos around active regions: a magnetohydrodynamic theory. Astrophys J 680:1457–1466. DOI 10.1086/587934. 0712.3578

Hanasoge SM (2009) A wave scattering theory of solar seismic power haloes. Astron Astrophys 503:595–599. DOI 10.1051/0004-6361/200912449. 0906.4671

Hanasoge SM (2014a) Full waveform inversion of solar interior flows. Astrophys J 797(23)

Hanasoge SM (2014b) Measurements and kernels for source-structure inversions in noise tomography. Geophys J Int 196:971–985. DOI 10.1093/gji/ggt411. 1310.0857

Hanasoge SM, Duvall TL Jr (2007) The solar acoustic simulator: % applications and results. Astron Nachr 328:319. DOI 10.1002/asna.200610737

Hanasoge SM, Duvall TL (2009) Subwavelength resolution imaging of the solar deep interior. Astrophys J 693:1678–1685. DOI 10.1088/0004-637X/693/2/1678. 0812.0119

Hanasoge SM, Larson TP (2008) Global effects of local sound-speed perturbations in the sun: a theoretical study. Sol Phys 251:91–100. DOI 10.1007/s11207-008-9208-6. 0711.1877

Hanasoge SM, Tromp J (2014) Full waveform inversion for time-distance helioseismology. Astrophys J 784(69):69. DOI 10.1088/0004-637X/784/1/69. 1401.7603

Hanasoge SM, Larsen RM, Duvall TL Jr, DeRosa ML, Hurlburt NE, Schou J, Roth M, Christensen-Dalsgaard J, Lele SK (2006) Computational acoustics in spherical geometry: steps toward validating helioseismology. Astrophys J 648: 1268–1275. DOI 10.1086/505927

Hanasoge SM, Duvall TL Jr, Couvidat S (2007) Validation of helioseismology through forward modeling: realization noise subtraction and kernels. Astrophys J 664:1234–1243. DOI 10.1086/519070

Hanasoge SM, Couvidat S, Rajaguru SP, Birch AC (2008) Impact of locally suppressed wave sources on helioseismic traveltimes. Mon Not R Astron Soc 391:1931–1939. DOI 10.1111/j.1365-2966.2008.14013.x. 0707.1369

Hanasoge SM, Duvall TL, DeRosa ML (2010a) Seismic constraints on interior solar convection. Astrophys J Lett 712:L98–L102. DOI 10.1088/2041-8205/712/1/L98. 1001.4508

Hanasoge SM, Komatitsch D, Gizon L (2010b) An absorbing boundary formulation for the stratified, linearized, ideal MHD equations based on an unsplit, convolutional perfectly matched layer. Astron Astrophys 522:A87. DOI 10.1051/0004-6361/201014345. 1003.0725

Hanasoge SM, Birch A, Gizon L, Tromp J (2011) The adjoint method applied to time-distance helioseismology. Astrophys J 738:100. DOI 10.1088/0004-637X/738/1/100. 1105.4263

Hanasoge SM, Duvall TL Jr, Sreenivasan KR (2012b) Anomalously weak solar convection. Proc Natl Acad Sci 109(30):11,928–11,932

Hartlep T, Miesch MS, Mansour NN (2008a) Wave propagation in the magnetic sun. arXiv e-prints 0805.0333

Hartlep T, Zhao J, Mansour NN, Kosovichev AG (2008b) Validating time-distance far-side imaging of solar active regions through numerical simulations. Astrophys J 689:1373–1378. DOI 10.1086/592721. 0805.0472

Harvey JW, Abdel-Gawad K, Ball W, Boxum B, Bull F, Cole J, Cole L, Colley S, Dowdney K, Drake R (1988) The GONG instrument. In: Rolfe EJ (ed) ESA, seismology of the sun and sun-like stars, pp 203–208. SEE N89-25819 19–92. http://adsabs.harvard.edu/abs/1988ESASP.286..203H

Hill F (1988) Rings and trumpets - three-dimensional power spectra of solar oscillations. Astrophys J 333:996–1013. DOI 10.1086/166807

Hill F, Fischer G, Grier J, Leibacher JW, Jones HB, Jones PP, Kupke R, Stebbins RT (1994) The global oscillation network group site survey. 1: data collection and analysis methods. Sol Phys 152:321–349. DOI 10.1007/BF00680443

Hirzberger J, Gizon L, Solanki SK, Duvall TL (2008) Structure and evolution of supergranulation from local helioseismology. Sol Phys 251:417–437. DOI 10.1007/s11207-008-9206-8

Hu FQ (2001) A stable, perfectly matched layer for linearized Euler equations in unsplit physical variables. J Comput Phys 173(2):455–480. DOI 10.1006/jcph.2001.6887. http://www.sciencedirect.com/science/article/B6WHY-45BC24W-1M/2/948d733fe6a2ba168e816194e9b481da

Hu FQ, Hussaini MY, Manthey JL (1996) Low-dissipation and low-dispersion Runge-Kutta schemes for computational acoustics. J Comput Phys 124(1): 177–191.   http://www.science-direct.com/science/article/B6WHY-45MGWBD-5S/1/8c70a06b8c29b3e947e1f0d1f2a65fc4

Hurlburt NE, Rucklidge AM (2000) Development of structure in pores and sunspots: flows around axisymmetric magnetic flux tubes. Mon Not R Astron Soc 314: 793–806

Igel H, Djikpéssé H, Tarantola A (1996) Waveform inversion of marine reflection seismograms for P impedance and Poisson's ratio. Geophys J Int 124: 363–371. DOI 10.1111/j.1365-246X.1996.tb07026.x

Jackiewicz J, Gizon L, Birch AC, Duvall TL Jr (2007) Time-distance helioseismology: sensitivity of f-mode travel times to flows. Astrophys J 671:1051–1064. DOI 10.1086/522914. 0708.3554

Jackiewicz J, Birch AC, Gizon L, Hanasoge SM, Hohage T, Ruffio JB, Švanda M (2012) Multichannel three-dimensional SOLA inversion for local helioseismology. Sol Phys 276:19–33. DOI 10.1007/s11207-011-9873-8. 1109.2712

Jain K, Komm RW, González Hernández I, Tripathy SC, Hill F (2012) Subsurface flows in and around active regions with rotating and non-rotating sunspots. Sol Phys 279:349–363. DOI 10.1007/s11207-012-0039-0. 1205.2356

Jameson A (1988) Aerodynamic design via control theory. J Sci Comput 3:233–260

Judge P, Kleint L, Uitenbroek H, Rempel M, Suematsu Y, Tsuneta S (2014) Photon mean free paths, scattering, and ever-increasing telescope resolution. arXiv e-prints 1409.7866

Khan A, Boschi L, Connolly J (2009) On mantle chemical and thermal heterogeneities and anisotropy as mapped by inversion of global surface wave data. J Geophys Res Solid Earth (1978–2012) 114(B9). DOI 10.1029/2009JB006399

Khomenko E, Collados M (2006) Numerical modeling of magnetohydrodynamic wave propagation and refraction in sunspots. Astrophys J 653:739–755. DOI 10.1086/507760

Khomenko E, Collados M (2008) Magnetohydrostatic sunspot models from deep subphotospheric to chromospheric layers. Astrophys J 689:1379–1387. DOI 10.1086/592681. 0808.3571

Khomenko E, Collados M (2009) Sunspot seismic halos generated by fast MHD wave refraction. arXiv e-prints 0905.3060

Kitchatinov LL, Rüdiger G (2005) Differential rotation and meridional flow in the solar convection zone and beneath. Astron Nachr 326:379–385. DOI 10.1002/asna.200510368. arXiv:astro-ph/0506239

Komatitsch D, Martin R (2007) An unsplit convolutional perfectly matched layer improved at grazing incidence for the seismic wave equation. Geophysics 72(5):SM155–SM167. DOI 10.1190/1.2757586

Komm R, González Hernández I, Hill F, Bogart R, Rabello-Soares MC, Haber D (2013) Subsurface meridional flow from HMI using the ring-diagram pipeline. Sol Phys 287:85–106. DOI 10.1007/s11207-012-0073-y

Korzennik SG, Rabello-Soares MC, Schou J (2004) On the determination of Michelson Doppler imager high-degree mode frequencies. Astrophys J 602:481–516. DOI 10.1086/381021. arXiv:astro-ph/0207371

Kosovichev AG, Duvall TL Jr (1997) Acoustic tomography of solar convective flows and structures. In: Pijpers FP, Christensen-Dalsgaard J, Rosenthal CS (eds) SCORe'96: solar convection and oscillations and their relationship. Astrophysics and space science library, vol 225. Springer, Berlin, pp 241–260

Kosovichev AG, Duvall TLJ, Scherrer PH (2000) Time-distance inversion methods and results - (invited review). Sol Phys 192:159–176

Kumar P, Basu S (2000) Source depth for solar P-modes. Astrophys J Lett 545: L65–L68. DOI 10.1086/317325. astro-ph/0006204

LeDimet FX, Talagrand O (1986) Variational algorithms for analysis and assimilation of meteorological observations: theoretical aspects. Tellus 38A:97–110

Leibacher JW, Stein RF (1971) A new description of the solar five-minute oscillation. Astrophys Lett 7:191–192

Leighton RB, Noyes RW, Simon GW (1962) Velocity fields in the solar atmosphere. I. preliminary report. Astrophys J 135:474. DOI 10.1086/147285

Lele SK (1992) Compact finite difference schemes with spectral-like resolution. J Comput Phys 103(1):16–42

Leroy B (1985) On the derivation of the energy flux of linear magnetohydrodynamic waves. Geophys Astrophys Fluid Dyn 32:123–133. DOI 10.1080/03091928508208781

Lighthill MJ (1952) On sound generated aerodynamically. I. general theory. Proc R Soc London Ser A 211(1107):564–587

Lindsey C, Braun DC (1997) Helioseismic holography. Astrophys J 485:895. DOI 10.1086/304445

Lions JL (1971) Optimal control of systems governed by partial differential equations. Springer, Berlin

Liu Q, Tromp J (2008) Finite-frequency sensitivity kernels for global seismic wave propagation based upon adjoint methods. Geophys J Int 174:265–286. DOI 10.1111/j.1365-246X.2008.03798.x

Lui C (2003) A numerical investigation of shock-associated noise. Ph.D. thesis, Stanford University

Luo Y, Modrak R, Tromp J (2013) Strategies in adjoint tomography. In: Freeden W, Nashed MZ, Sonar T (eds) Handbook of geomathematics, 2nd edn. Springer, Berlin

Lynden-Bell D, Ostriker JP (1967) On the stability of differentially rotating bodies. Mon Not R Astron Soc 136:293

Meza-Fajardo KC, Papageorgiou AS (2008) A nonconvolutional, split-field, perfectly matched layer for wave propagation in isotropic and anisotropic elastic media: stability analysis. Bull Seismol Soc Am 98(4):1811–1836. http://www. bssaonline.org/cgi/content/abstract/98/4/1811

Miesch MS, Hindman BW (2011) Gyroscopic pumping in the solar near-surface shear layer. Astrophys J 743:79. DOI 10.1088/0004-637X/743/1/79. 1106.4107

Miesch MS, Featherstone NA, Rempel M, Trampedach R (2012) On the amplitude of convective velocities in the deep solar interior. Astrophys J 757:128. DOI 10. 1088/0004-637X/757/2/128. 1205.1530

Moradi H, Cally PS (2014) Sensitivity of helioseismic travel times to the imposition of a Lorentz force limiter in computational helioseismology. Astrophys J Lett 782:L26. DOI 10.1088/2041-8205/782/2/L26. 1401.5518

Moradi H, Hanasoge SM, Cally PS (2009) Numerical models of travel-time inhomogeneities in sunspots. Astrophys J Lett 690:L72–L75. DOI 10.1088/0004-637X/ 690/1/L72. 0808.3628

Moradi H, Baldner C, Birch AC, Braun DC, Cameron RH, Duvall TL, Gizon L, Haber D, Hanasoge SM, Hindman BW, Jackiewicz J, Khomenko E, Komm R, Rajaguru P, Rempel M, Roth M, Schlichenmaier R, Schunker H, Spruit HC, Strassmeier KG, Thompson MJ, Zharkov S (2010) Modeling the subsurface structure of sunspots. Sol Phys 267:1–62. DOI 10.1007/s11207-010-9630-4. 0912.4982

Nordlund Å, Stein RF, Asplund M (2009) Solar surface convection. Living Rev Sol Phys 6:2

Noyes RW, Weiss NO, Vaughan AH (1984) The relation between stellar rotation rate and activity cycle periods. Astrophys J 287:769–773. DOI 10.1086/162735

Orszag SA (1971) On the elimination of aliasing in finite difference schemes by filtering high-wavenumber components. J Atmos Sci 28:1074

Parchevsky KV, Kosovichev AG (2007) Three-dimensional numerical simulations of the acoustic wave field in the upper convection zone of the sun. Astrophys J 666:547–558. DOI 10.1086/520108. arXiv:astro-ph/0612364

Parchevsky KV, Zhao J, Kosovichev AG (2008) Influence of nonuniform distribution of acoustic wavefield strength on time-distance helioseismology measurements. Astrophys J 678:1498–1504. DOI 10.1086/533495. arXiv:0802.3866

Parker EN (1979) Cosmical magnetic fields: their origin and their activity. Clarendon Press, Oxford; Oxford University Press, New York, 858 p.

Proffitt CR, Michaud G (1990) Gravitational settling in solar models. In: Bulletin of the American Astronomical Society. Bulletin of the American Astronomical Society, vol 22, p 1198. http://adsabs.harvard.edu/abs/1990BAAS...22Q1198P

Pulkkinen T (2007) Space weather: terrestrial perspective. Living Rev Sol Phys 4(1). http://www.livingreviews.org/lrsp-2007-1; http://adsabs.harvard.edu/ abs/2007LRSP....4....1P

Ravaut C, Operto S, Improta L, Virieux J, Herrero A, Dell'Aversana P (2004) Multiscale imaging of complex structures from multifold wide-aperture seismic data by frequency-domain full-waveform tomography: application to a thrust belt. Geophys J Int 159:1032–1056. DOI 10.1111/j.1365-246X.2004.02442.x

Rempel M (2005) Solar differential rotation and meridional flow: the role of a subadiabatic tachocline for the Taylor-Proudman balance. Astrophys J 622: 1320–1332. DOI 10.1086/428282. arXiv:astro-ph/0604451

Rempel M, Schüssler M, Knölker M (2009) Radiative magnetohydrodynamic simulation of sunspot structure. Astrophys J 691:640–649. DOI 10.1088/0004-637X/691/1/640. 0808.3294

Rhodes EJ Jr, Ulrich RK, Simon GW (1977) Observations of nonradial p-mode oscillations on the sun. Astrophys J 218:901–919. DOI 10.1086/155745

Rickers F, Fichtner A, Trampert J (2013) The Iceland–Jan Mayen plume system and its impact on mantle dynamics in the North Atlantic region: evidence from full-waveform inversion. Earth Planet Sci Lett 367:39–51

Roden JA, Gedney SD (2000) Convolution PML (CPML): an efficient FDTD implementation of the CFS-PML for arbitrary media. Microw Opt Technol Lett 27(5):334–339

Rosenwald RD, Rabaey GF (1991) Application of the continuous orthonormalization and adjoint methods to the computation of solar eigenfrequencies and eigenfrequency sensitivities. Astrophys J Suppl 77:97–117. DOI 10.1086/191600

Scherrer PH, Bogart RS, Bush RI, Hoeksema JT, Kosovichev AG, Schou J, Rosenberg W, Springer L, Tarbell TD, Title A, Wolfson CJ, Zayer I, MDI Engineering Team (1995) The solar oscillations investigation - Michelson Doppler imager. Sol Phys 162:129–188. DOI 10.1007/BF00733429

Schlüter A, Temesváry S (1958) The internal constitution of sunspots. In: Lehnert B (ed) Electromagnetic phenomena in cosmical physics, IAU symposium, vol 6, p 263

Schou J, Antia HM, Basu S, Bogart RS, Bush RI, Chitre SM, Christensen-Dalsgaard J, di Mauro MP, Dziembowski WA, Eff-Darwich A, Gough DO, Haber DA, Hoeksema JT, Howe R, Korzennik SG, Kosovichev AG, Larsen RM, Pijpers FP, Scherrer PH, Sekii T, Tarbell TD, Title AM, Thompson MJ, Toomre J (1998) Helioseismic studies of differential rotation in the solar envelope by the solar oscillations investigation using the Michelson Doppler imager. Astrophys J 505: 390–417. DOI 10.1086/306146

Schou J, Scherrer PH, Bush RI, Wachter R, Couvidat S, Rabello-Soares MC, Bogart RS, Hoeksema JT, Liu Y, Duvall TL, Akin DJ, Allard BA, Miles JW, Rairden R, Shine RA, Tarbell TD, Title AM, Wolfson CJ, Elmore DF, Norton AA, Tomczyk S (2012) Design and ground calibration of the helioseismic and magnetic imager (HMI) instrument on the solar dynamics observatory (SDO). Sol Phys 275: 229–259. DOI 10.1007/s11207-011-9842-2

Schrijver CJ, Zwaan C (2000) Solar and stellar magnetic activity. Cambridge astrophysics series, vol 34. Cambridge University, Cambridge. http://adsabs.harvard.edu/abs/2000ssma.book.....S

Schrijver CJ, Hagenaar HJ, Title AM (1997) On the patterns of the solar granulation and supergranulation. Astrophys J 475:328. DOI 10.1086/303528

Schunker H, Braun DC, Cally PS, Lindsey C (2005) The local helioseismology of inclined magnetic fields and the showerglass effect. Astrophys J Lett 621: L149–L152. DOI 10.1086/429290

Schunker H, Cameron RH, Gizon L, Moradi H (2011) Constructing and characterising solar structure models for computational helioseismology. Sol Phys 271:1–26. DOI 10.1007/s11207-011-9790-x. 1105.0219

Shapiro NM, Campillo M (2004) Emergence of broadband Rayleigh waves from correlations of the ambient seismic noise. Geophys Res Lett 31:L07,614. DOI 10.1029/2004GL019491

Shelyag S, Erdélyi R, Thompson MJ (2006) Forward modeling of acoustic wave propagation in the quiet solar subphotosphere. Astrophys J 651:576–583. DOI 10.1086/507463

Shelyag S, Zharkov S, Fedun V, Erdélyi R, Thompson MJ (2009) Acoustic wave propagation in the solar sub-photosphere with localised magnetic field concentration: effect of magnetic tension. Astron Astrophys 501:735–743. DOI 10.1051/0004-6361/200911709. 0901.3680

Sirgue L, Pratt RG (2004) Efficient waveform inversion and imaging: a strategy for selecting temporal frequencies. Geophysics 69:231–248

Snieder R (2004) Extracting the Green's function from the correlation of coda waves: a derivation based on stationary phase. Phys Rev E 69(4):046610. DOI 10.1103/PhysRevE.69.046610

Stein RF, Nordlund Å (2000) Realistic solar convection simulations. Sol Phys 192:91–108. DOI 10.1023/A:1005260918443

Stix M (2004) The Sun: and introduction. Springer, Berlin

Švanda M (2012) Inversions for average supergranular flows using finite-frequency kernels. Astrophys J Lett 759:L29. DOI 10.1088/2041-8205/759/2/L29. 1209.6147

Švanda M, Gizon L, Hanasoge SM, Ustyugov SD (2011) Validated helioseismic inversions for 3D vector flows. Astron Astrophys 530:A148. DOI 10.1051/0004-6361/201016426. 1104.4083

Swisdak M, Zweibel E (1999) Effects of large-scale convection on p-mode frequencies. Astrophys J 512:442–453. DOI 10.1086/306764. arXiv:astro-ph/9809135

Talagrand O, Courtier P (1987) Variational assimilation of meteorological observations with the adjoint vorticity equation. I: theory. Q J R Meteorol Soc 113:1311–1328. DOI 10.1256/smsqj.47811

Tape C, Liu Q, Tromp J (2007) Finite-frequency tomography using adjoint methods — methodology and examples using membrane surface waves. Geophys J Int 168:1105–1129

Tape C, Liu Q, Maggi A, Tromp J (2009) Adjoint tomography of the Southern California crust. Science 325(5943):988–992. DOI 10.1126/science.1175298. http://www.sciencemag.org/cgi/content/abstract/325/5943/988. http://www.sciencemag.org/cgi/reprint/325/5943/988.pdf

Tarantola A (1984) Linearized inversion of seismic reflection data. Geophys Prospect 32:998–1015

Thompson KW (1990) Time-dependent boundary conditions for hyperbolic systems. J Comput Phys 89(2):439–461

Thoul AA, Bahcall JN, Loeb A (1994) Element diffusion in the solar interior. Astrophys J 421:828–842. DOI 10.1086/173695. astro-ph/9304005

Tóth G (2000) The divb=0 constraint in shock-capturing magnetohydrodynamics codes. J Comput Phys 161(2):605–652 http://www.sciencedirect.com/science/article/B6WHY-45FC8HX-47/2/61db392c9a72d394a62a02958881b046

Tromp J, Tape C, Liu Q (2005) Seismic tomography, adjoint methods, time reversal and banana-doughnut kernels. Geophys J Int 160:195–216. DOI 10.1111/j.1365-246X.2004.02453.x

Tromp J, Luo Y, Hanasoge S, Peter D (2010) Noise cross-correlation sensitivity kernels. Geophys J Int 183:791–819, DOI 10.1111/j.1365-246X.2010.04721.x

Ulrich RK (1970) The five-minute oscillations on the solar surface. Astrophys J 162:993. DOI 10.1086/150731

Vichnevetsky R, Bowles JB (1982) Fourier analysis of numerical approximations of hyperbolic equations, vol 5. Society for Industrial and Applied Mathematics, Philadelphia

Winton SC, Rappaport CM (2000) Specifying PML conductivities by considering numerical reflection dependencies. IEEE Trans Antennas Propag 48(7):1055–1063

Woodard M (2014) Detectability of large-scale solar subsurface flows. Sol Phys 289:1085–1100. DOI 10.1007/s11207-013-0386-5

Woodard M, Schou J, Birch AC, Larson TP (2013) Global-oscillation eigenfunction measurements of solar meridional flow. Sol Phys 287:129–147. DOI 10.1007/s11207-012-0075-9

Woodard MF (1997) Implications of localized, acoustic absorption for heliotomographic analysis of sunspots. Astrophys J 485:890–894. DOI 10.1086/304468

Woodard MF (2007) Probing supergranular flow in the solar interior. Astrophys J 668:1189–1195. DOI 10.1086/521391

Zhao J, Kosovichev AG (2003) On the inference of supergranular flows by timedistance helioseismology. In: Sawaya-Lacoste H (ed) Proceedings of SOHO 12/GONG+ 2002. Local and global helioseismology: the present and future, Big Bear Lake, 27 October-1 November 2002. ESA SP-517. ESA, Noordwijk, pp 417–420. ISBN 92-9092-827-1. http://adsabs.harvard.edu/abs/2003ESASP.517..417Z

Zhao J, Kosovichev AG, Duvall TL Jr (2001) Investigation of mass flows beneath a sunspot by time-distance helioseismology. Astrophys J 557:384–388. DOI 10.1086/321491

Zhao J, Hartlep T, Kosovichev AG, Mansour NN (2009) Imaging the solar tachocline by time-distance helioseismology. Astrophys J 702:1150–1156. DOI 10.1088/0004-637X/702/2/1150. 0907.2118

Zhao J, Bogart RS, Kosovichev AG, Duvall TL Jr, Hartlep T (2013) Detection of equatorward meridional flow and evidence of double-cell meridional circulation inside the sun. Astrophys J Lett 774:L29. DOI 10.1088/2041-8205/774/2/L29. 1307.8422

Zhu H, Luo Y, Nissen-Meyer T, Morency C, Tromp J (2009) Elastic Imaging and time-lapse migration based on adjoint methods. Geophysics 74(6):167

Zhu H, Bozdağ E, Duffy TS, Tromp J (2013) Seismic attenuation beneath Europe and the North Atlantic: implications for water in the mantle. Earth Planet Sci Lett 381(0):1–11. Doi http://dx.doi.org/10.1016/j.epsl.2013.08.030. http://www.sciencedirect.com/science/article/pii/S0012821X13004676

Printed in the United States
By Bookmasters